Signals in the Noise

Security, Technology, and the Hidden Paterns of Modern Risk

Timothy E. Reed

Book Cover Design
Dian Azura

ICE STATION
ZEBRA
PRESS

Signals in the Noise

For permission requests, contact: Ice Station Zebra Press

In partnership with: The Reed Group

Pittsburgh, Pennsylvania

Email: info@isz.one

Website: www.signalsinthenoisebook.com

Library of Congress Control Number: *[to be assigned]*

ISBN:

979-8-9936676-0-7 (e-book), 979-8-9936676-1-4 (paperback)

979-8-9936676-2-1 (hardcover) 979-8-9936676-3-8 (audiobook)

Cover design by Dian Azura

Printed in the United States of America

10 9 8 7 6 5 4 3 2 1

First Edition

"Security is a process, not a product."™ and Ice Station Zebra® are trademarks of Timothy E. Reed.

Foreword

The modern enterprise runs on data but struggles with clarity. Information pours through every system, every sensor, every interaction. Metrics, dashboards, and alerts create the impression of control, yet leaders often find themselves staring at more inputs and fewer answers. The challenge is not a lack of visibility. It is knowing what matters.

We've reached a point where velocity outpaces understanding. Decisions are made in compressed cycles with limited context. Automation accelerates everything, including errors. Teams move fast because the environment demands it, not because the conditions support it. The result is a widening gap between what leaders *see* and what they can *trust*.

This book is about closing that gap.

Clarity is no longer a passive trait. It must be engineered, protected, and practiced. The organizations that excel in the decade ahead will not be the ones with the most data. They will be the ones with the discipline to interpret the signal in a world built to generate noise.

Across these chapters, you will find a framework for navigating the modern risk landscape: how information can be distorted, how algorithms shape perception, how ambiguity can be weaponized, and how the human mind remains essential in a system increasingly defined by automation. You will see why traditional models for risk, intelligence, and organizational awareness no longer fit the world we've built, and what must replace them.

The goal is simple: give leaders a sharper way to think.

If we are to thrive in an era of rapid change, we need more than technology. We need judgment. We need structure. We need intentional awareness rather than reactive awareness. Clarity is not a luxury. It is an operational requirement.

This book shows how to achieve it.

To my wife, Jiamin, thank you for the steady push to follow through and the constant belief that made this real.

To our son, Declan, I am proud of the young man you are becoming. Being your father is the best work of my life. Follow your heart, stay curious, and keep growing into the remarkable person you already are.

To those who shaped my professional path—Mother, JDB, RD, KS, GM, JT, and Haz—your guidance, candor, and patience opened doors I could not have found on my own. For those of you who have provided editorial feedback, BJ, KR, and others, I couldn't have done it without your candid feedback.

Nick, the world isn't the same without you. I would have loved to spar with you as I build this out.

To my parents and extended family: from rural Western New York to nearly fifteen years overseas, your foundation gave me the courage to leap and the strength to land. Everything here stands on what you gave me.

Prelude

The Fog of Signal

Everything appears urgent. Everything demands attention. Yet urgency is not the same as importance. By the time a leader opens their laptop each morning, dashboards, alerts, and messages are already competing for priority.

Every system we rely on—commercial, social, technological—is now designed to generate continuous noise. Notifications, analyses, dashboards, and automated summaries create the impression that we're fully informed. They fragment attention. The more we attempt to monitor, the less we understand.

They're not overwhelmed because the world is unknowable, but because the system is tuned for continuous stimulation. This isn't a failure of intelligence or effort. It's a structural issue. When data grows faster than comprehension, clarity becomes an active skill. The ability to pause, examine the environment, and identify the relevant pattern is what separates effective leadership from frantic reaction.

This book builds on a simple idea: Awareness requires intention. One clarifying question in a morning briefing often reveals more than a dashboard of metrics.

The chapters that follow examine how modern systems change how we perceive risk, how algorithmic filters shape perception, how ambiguity is used to influence decision-making, and how humans can regain stability in an information ecosystem not built for them. You will find examples from industry, security, and everyday life, but the purpose is not to catalog threats. It's to show a way to think— one that lets you navigate complexity without being consumed by it.

The world will continue to accelerate. Data will continue to multiply. Systems will continue to automate decisions that were previously left to humans. Automation can approve a transaction instantly, but only a human sees the context that makes it suspicious. None of that is inherently dangerous. What becomes dangerous is failing to recognize how these forces alter judgment.

Clarity is not found by chasing every input. It comes from understanding which patterns deserve attention and which are artifacts of noise. The skill is not to see more, but to know what you are seeing. That distinction is what turns information into advantage.

This is your guide to making that distinction.

Chapter 1

The Collapse of Signal

The Overload Condition

At 09:23, the system went into alarm.

A camera feed from a distribution site in Texas. Seconds later, it froze, then multiplied across eight identical frames tiled across the wall—chat alerts stacked in vertical panic. The AI triage bot flagged a possible anomaly. Another dashboard lit up in sympathetic hysteria.

The room reacted like a reflex arc, muscles before thought: more screens, more data, more movement. Chairs scraped, and fingers flew before anyone had time to think. Within minutes, the analysts were no longer solving a problem; they were managing the noise generated by the response.

. . .

This is the new rhythm of operational life, the illusion of control at the speed of confusion. Command centers are built with more tools, more screens, and more data. The architecture looks impressive, but the fundamental question remains. Does any of this serve its intended purpose?

We have built systems that simulate awareness so completely they obscure it. Every ping feels like progress. Every new data layer promises mastery. Yet attention is finite, and velocity is not comprehension.

The irony is almost mathematical. Systems designed to make us more informed have made us functionally blind. Information has achieved escape velocity. It is no longer something we collect and interpret; it is a self-sustaining weather pattern that threatens our survival.

In the modern command center, dashboards never sleep. Algorithms scrape, correlate, and escalate. Metrics generate meaning the way clouds create storms. Operators, human or otherwise, stand inside the data hurricane and call it visibility.

The overload condition does not present as chaos. It presents as productivity. The dashboard glows—the analytics stream updates by the second. Meetings proliferate. Reports lengthen. Somewhere in the static, a quiet voice insists that you are on top of this.

· · ·

Until you are not.

When everything is urgent, nothing is essential. When everything is visible, nothing is truly seen.

The cost of infinite information is not ignorance. It is a distortion. A team may have more data than ever, yet still miss the one signal that actually matters because every metric is lit equally brightly.

The Inversion of Advantage

The dynamics of overload are not new. For most of human history, power was defined by access, whether it was secrets, sources, or knowledge withheld from others: a monarch's cipher, a general's map, an analyst's file. The advantage was asymmetrical.

That advantage is gone.

Access has become universal, instantaneous, and overwhelming. Everyone sees roughly the same feeds, reads the same reports, and scrapes the same digital commons. The differentiator is no longer who knows, but who discerns.

What was once a moat has become a floodplain. Information

flows everywhere, and with it comes a new problem: the aristocracy of noise.

Corporations, governments, and individuals now experience what intelligence agencies faced at the dawn of the satellite era: the vertigo of omniscience. The more you can see, the less you know. Every organization now lives in the orbital regime, surrounded by data and unable to escape its gravity.

Strategic plans drown under performance dashboards. Analysts chase reflections of reflections. Predictive models confirm the last model's conclusions, creating a loop of reassurance rather than genuine insight. We were trained to treat data as a source of illumination. Instead, it has become glaring.

The competitive edge, once defined by information superiority, has inverted. The organizations that move fastest often see the least. Speed amplifies error. Connectivity multiplies confusion. The rarest capability is not real-time data; it is real-time understanding.

That inversion reshapes everything: how we lead, how we decide, and how we define secure.

A company's survival no longer depends on its ability to collect more information than its rivals. It depends on its ability to interpret less, better. The advantage belongs to those who can navigate the flood of information without being swept away.

The Lost Art of Discernment

Security was never meant to be a product. It was, and remains, a process. That distinction matters more now than ever. A product promises finality: buy this, install that, achieve safety. A process demands engagement, observation, interpretation, adaptation, and repetition. One ends in complacency. The other cultivates awareness. The same principle governs intelligence, leadership, and perception.

Discernment is not delivered through an interface. It takes shape in the brief pause before reacting to an alert, the moment when a leader asks what the system may not be seeing. It is forged through disciplined observation, contextual empathy, and the humility to admit uncertainty. True discernment begins with a simple act: slowing down and letting patterns reveal themselves rather than forcing them to appear.

In practice, this means retraining both people and systems to privilege meaning over motion. It means designing workflows that absorb ambiguity instead of trying to eradicate it. Restoring reflection to the response tempo might mean building deliberate pauses into automated escalation paths, giving analysts space to evaluate rather than react.

. . .

The lesson is not new. During the Second World War, the codebreakers at Bletchley Park faced a problem that could not be solved through force or haste. The German Enigma cipher seemed impenetrable, not because it was impossible to break, but because its patterns were buried beneath immense noise. The analysts did not rush to conclusions. They collected routine transmissions — weather reports, repeated salutations, mundane messages — until faint structures surfaced. Meaning emerged not from speed, but from sustained attention.

Accurate pattern recognition depends on volume and patience. Rushing destroys the signal. Automation can accelerate awareness, but it cannot create it. AI can summarize context, but it cannot sense it.

The lost art of discernment begins where automation ends. It is the moment when data becomes understanding and understanding becomes a decision. That is the purpose of this book: to recover the habits of mind and design that enable signals to resurface, rebuilding clarity as an operational capability.

Ultimately, a signal collapse is not a technological failure. It is a human one.

And the way back begins, as all sound security does, as a process, not a product.

Chapter 2
Weaponized Ambiguity

The Cognitive Battlefield

The modern battlefield is no longer territorial. It is cognitive. Its terrain is attention, and its weapon is interpretation. A trending clip, a fragment of context, a viral caption can now do what artillery once did: shift belief. No ordinance, no borders, no uniforms, just streams of perception competing for control of what seems true.

In earlier eras, the power to shape perception belonged to those who controlled land or armies. Today, it belongs to those who control narrative velocity, the speed at which meaning spreads and solidifies. The decisive high ground is no longer a hilltop. It is the first page of a search result, the feed that defines relevance, or the algorithmic lens that filters reality before we see it. The shift is not new; it is only more complete, woven into every tap, swipe, and automated feed we consume.

. . .

During the Cold War, both the United States and the Soviet Union waged psychological and informational campaigns alongside their military ones. Radio Free Europe, Soviet disinformation, and carefully manipulated images were not side operations. They were central theaters of conflict. The goal was not only to inform or mislead, but to condition belief. Even then, the battlefield was drifting from the physical to the perceptual.

A headline could rally a population faster than a convoy. And as perception became the terrain, people themselves became part of it. The twenty-first century has turned every citizen into both participant and target. Social platforms serve as real-time battlegrounds where perception is shaped algorithmically, not deliberatively. A well-timed deepfake can move markets. A viral post can move armies. A bot swarm can fabricate consensus. Influence has been democratized. Discernment has not.

The examples are familiar. The Arab Spring showed how digital networks could mobilize populations faster than regimes could react, turning attention into an accelerant of political change. The 2016 U.S. election and its aftermath revealed how external actors could weaponize emotion to fracture civic trust. In Ukraine, satellite imagery, Telegram channels, and drone footage coexist with deliberate falsehoods. War now occurs in trenches, timelines, and feeds simultaneously, where a repost can travel farther than a regiment.

The private sector is also drawn into this cognitive theater. Corporations face misinformation campaigns that can devalue stock within hours. Artificial intelligence generates persuasive but synthetic "evidence." Synthetic personas debate real ones. Every institution, from governments to brands, must defend not only its infrastructure, but its credibility – because credibility is now the easiest thing to erode.

The new battle for truth is fought in milliseconds and measured in impressions. It does not seek to occupy territory, but to erode confidence. Confidence in the news. Confidence in institutions. Confidence in each other. The strategic objective is not victory. It is confusion, because confusion slows judgment – and slowed judgment shapes outcomes.

Cognitive warfare rewards those who can discern without reacting, who can see patterns without being pulled into their gravity. The first defence is awareness: noticing when a narrative feels engineered. The second is composure: resisting the impulse to react instantly, especially when speed is the pressure point. In this domain, clarity itself is a form of armor.

A single well-placed ambiguity, amplified through millions of automated agents, can distort perception across an entire population. It does not need to be false. It just needs to linger.

Misinformation is not simply a falsehood. It is friction – the tiny hesitation when a headline contradicts a video, and the mind stutters between them; that drag between what is seen and what is believed. It does not destroy the truth. It dilutes it. A fragment of narrative lifted out of context, and coherence unravels.

Modern influence works through disorientation, not persuasion. Not by convincing you, but by exhausting you. Every scroll, every refresh, every breaking update extends the front line. The battle is fought less over what happened and more over what it means, and who gets to define that meaning first. The human brain, designed for pattern and purpose, becomes the most exploitable system in the loop.

Ambiguity as a Service

Ambiguity has become an industry.

Algorithms amplify confusion because confusion is profitable. Platforms reward engagement, not accuracy – outrage scales. Truth does not. In most systems, attention spikes long before understanding can catch up.

. . .

In the architecture of modern media, clarity is rarely rewarded. Often, it is penalized. The more uncertain the user, the longer they stay. The longer they stay, the more data they generate. In an economy driven by attention metrics, confusion is capital. Every hesitation, every rewatch, every moment of doubt becomes monetizing behavior. What began as a side effect of engagement design has matured into a strategy.

Every click becomes an endorsement of entropy. Each interaction tells the system that the signal, even if inaccurate, should be repeated. Every share becomes a subsidy for noise. The systems do not care what you believe. They care that you keep thinking something. Engagement is the metric. Ambiguity is the fuel that keeps the engine running.

These same design choices bleed into security and intelligence work. The psychological levers that keep users scrolling now influence how leaders and analysts process information. Rapid response is treated as a virtue. Nuance is treated as a liability.

Bias is no longer just ideological. It is algorithmic. The feed learns your fears and reflects them to you, refining its predictions with every interaction. The result is a closed perceptual circuit. A personalized echo chamber that feels objective precisely because it feels tailored. The illusion of insight replaces actual understanding.

· · ·

For security programs, this is not just a cultural concern. It is an operational threat. Analysts work in an environment saturated with algorithmic reinforcement, where open-source intelligence, social sentiment, and real-time indicators are all shaped by unseen layers of curation. Machine bias becomes human bias. Human bias trains the machine in return. Each crank turn strengthens the flywheel.

False positives proliferate. Analysts prioritize trends over what matters—situational awareness fragments into microbursts of attention. In crisis response, this cognitive distortion can lead to real harm, rushed decisions, and loss of context due to velocity.

The feedback loop reaches leadership as well. Decision-makers crave dashboards, metrics, and trend indicators that appear to reflect the world, but often measure engagement rather than reality. Security dashboards are beginning to resemble social feeds: colorful, dynamic, and subtly misleading. Metrics create the illusion of control while deepening dependency.

"Facts are stubborn things, but statistics are more pliable." - Mark Twain

You feel informed, connected, and responsive, while the underlying structure feeds on that sensation. The same dynamics scale far beyond the individual. Financial markets respond to rumors before analysis. News cycles chase perception over verification. Intelligence assessments risk contamination by the same forces that drive viral content. Speed becomes synonymous with relevance. Verification becomes a luxury.

The truth has not disappeared. It simply no longer pays as well. In an attention-driven system, accuracy slows the loop. In a system optimized for attention, accuracy is friction. The signal must compete not only with noise but also with the algorithms designed to amplify it.

The new discipline of intelligence is not only to find the truth. It is to defend the conditions under which truth can still be seen.

The Geometry of Confusion

At scale, ambiguity behaves like geometry. It shapes movement and refracts meaning. Its influence becomes visible only when you step back far enough to see the pattern. In any crisis, the informational terrain warps first. The earliest distortions are often the smallest.

A single ambiguous phrase, an image without context, a misquoted source, or a synthetic headline creates a gravitational pull. Analysts pivot to investigate. Users debate. Algorithms amplify the tension between interpretations. Within minutes, ambiguity becomes architecture, a self-sustaining structure around which entire narratives form. It endures because each interpretation reinforces the next.

This is how control works now. Not by dictating what people think, but by defining the frame in which thinking occurs. Weaponized ambiguity does not aim for consensus. It aims for paralysis. It makes decisive action impossible because every path feels compromised by doubt.

From a security perspective, this is not just disinformation. It is a denial of cognition. It disrupts the capacity to form a coherent interpretation. The objective is not to persuade or deceive, but to saturate the environment with enough interpretive noise that no clear decision can emerge.

In that environment, speed is not an advantage. It is an accelerant. The faster an organization reacts, the more likely it is to respond to the wrong signal. Velocity without verification becomes vulnerability.

Historical Precedent: The Fog of 1941

In the weeks before the attack on Pearl Harbor, American intelligence intercepted fragments of Japanese diplomatic and naval communications. The signals were partial and ambiguous, and they were interpreted through existing assumptions. This led to competing readings of the same data. Analysts debated the meaning while field commanders waited for clarity. Some fragments pointed toward Southeast

Asia. Others hinted at the Pacific. None of the pieces was false, but the pattern remained hidden beneath uncertainty.

Ambiguity effectively paralyzed cognition. Decision-makers faced an information geometry that bent judgment inward. Too much data to ignore. Too little clarity to act. When the attack came, it was not intelligence that failed. It was an interpretation. The signals were seen, but their meaning was refracted through competing narratives.

This pattern has replayed across decades. The Cuban Missile Crisis, the Arab Spring, the early stages of COVID-19, and the 9/11 attacks all began as battles for informational coherence before they became battles for control. Only after careful, retrospective analysis did the signal appear obvious.

In the modern era, the exact geometry unfolds algorithmically. An unverified post can shape policy discussions before any analyst verifies its source. A misinterpreted video clip can provoke diplomatic fallout in minutes. The terrain of decision-making now shifts in real time, bending under the weight of collective reaction.

The necessary skill is not raw speed. It is calibration. The leaders who prevail in this environment are the ones who wait long enough for the noise to settle into a pattern. Clarity, in an age of weaponized ambiguity, is not a state. It is a practice. One built on waiting, not reacting.

The Discipline of Clarity

You cannot fight ambiguity with volume. You fight it with structure, precision, and patience.

The antidote to ambiguity is discipline. Not the rigidity of control, but the rigor of clarity. Clarity is not simplicity. It is the deliberate reduction of distortion. It is the act of maintaining fidelity to meaning when everything around you is optimized for distraction.

This demands a shift in how organizations relate to information. We need to move from consuming data to curating sense, from reacting to interpreting, from monitoring to discerning.

In practice, this means designing systems that penalize reflex and reward verification. It requires slowing the interpretive cycle. It means embedding context loops into automation so that every output carries its lineage. Where the data came from, how it was transformed, and what assumptions it holds means teaching analysts and leaders to tolerate ambiguity without surrendering to it, and to stand in uncertainty long enough for clarity to emerge.

In an age of automated confusion, clarity becomes a subversive act. To communicate clearly is to resist. To interpret deliberately is to defend. To see with intent is to reclaim cognition from the machines built to harvest it.

The future of intelligence and leadership does not depend on more data or faster systems. It depends on rediscovering a human skill we once took for granted: the ability to think clearly under fire.

Historical Example: The Cuban Missile Crisis (1962)

Few moments in modern history demonstrate the discipline of clarity under pressure as clearly as the thirteen days of the Cuban Missile Crisis.

When U.S. reconnaissance flights revealed Soviet missile sites under construction in Cuba, the informational environment collapsed into chaos. Signals multiplied faster than they could be interpreted. Intelligence agencies, military leaders, and politicians each interpreted the same data through different lenses. Some advocated immediate airstrikes. Others urged diplomacy. Every hour carried existential weight.

President John F. Kennedy's next move was a masterclass in structured discernment. Instead of reacting to the flood of ambiguous signals, he slowed the tempo. He created space for evaluation. He formed the Executive Committee of the National Security Council, a group tasked not with producing more information but with achieving a more straightforward interpretation.

Meetings were held without expecting immediate decisions. Arguments were encouraged. Dissent was documented. Assumptions were surfaced. Analysts were pushed to trace the lineage of their intelligence. How confident were they in their sources? What did they not know? Which biases colored their assessments? Each question clarified the edges of uncertainty.

Kennedy delayed action until clarity improved. The result was a calibrated, deliberate response, a naval quarantine that projected strength without triggering escalation. The world stepped back from nuclear war not because America had superior weapons or faster intelligence, but because its leadership had the discipline to interpret slowly.

The crisis showed that clarity, not speed, preserves stability.

Modern Parallel: The 2020 SolarWinds Cyber Breach

Nearly six decades later, the same principle reappeared in the digital realm.

When U.S. agencies discovered that SolarWinds Orion software had been compromised, early reporting described what appeared to be a total network collapse. Across government and industry, the instinct was to respond immediately by shutting down systems, revoking credentials, and purging code.

Teams that practiced restraint uncovered a different reality. The intrusion was far more selective than initial reports suggested. The attackers had implanted targeted backdoors that could only be mapped through methodical, patient analysis. That patience was essential. Acting too quickly would have destroyed the forensic evidence needed to understand the scope and intent of the attack.

The most effective responses came from teams that embraced the discipline of clarity. They slowed down, mapped context, and demanded verification before action. Their success rested less on the speed of detection and more on the precision of understanding.

Across centuries and domains, from missile crises to malware, the lesson remains the same. Velocity is seductive. Clarity is sovereign.

The leaders who will navigate the next age of noise are not the ones who process the most data. They are the ones who can still think slowly enough to see truly.

The Quiet Flag

The security operations center was quiet except for the soft hiss of dashboards. On a Wednesday, just after six in the morning, a low-confidence alert appeared in the identity provider feed, indicating an unusual login attempt. One contractor account attempted to log in from Bucharest using a consumer ISP. The risk engine scored it as a curiosity, not a crisis.

The contractor's profile listed the United States as home base, and a Salesforce role created two days earlier by HR. The analyst on duty, Lena, noticed a mismatch between a brand-new account and a foreign IP address that did not match the onboarding form.

Signal confusion. The indicators pointed in different directions. The case could go two ways. It might be legitimate travel. It might be an attacker testing a weak initial password before the user sets multifactor. Lena checked the HR system. The contractor's start date was yesterday. The travel field was blank.

She checked the ticket that created the account. It mentioned a third-party agency and a Product Ops hiring manager. Slack messages showed a welcome note posted late the previous night. Nothing mentioned travel.

· · ·

The identity product's risk rank left the alert below louder events in the queue, but the story felt brittle. New account. Foreign IP. Consumer network. No travel note.

Lena set a thirty-minute decision window. She posted a short entry in the incident log. She triggered a step-up challenge for the account and paused automatic group entitlements in the identity tool. She messaged the hiring manager and the agency contact and marked both as priority.

While the clock was running, she looked for corroborating evidence. Something independent of the login trail. She found the contractor's personal LinkedIn post from the weekend. It showed a flight to Europe for a family emergency and a note that they would be working irregular hours for the first week.

The hiring manager replied within the 10-minute window. Yes, the contractor was overseas temporarily. No, they had not yet enrolled their hardware security key. The agency admitted they had reused the temporary password in an email chain to the contractor because the portal was throwing errors.

The outcome and lesson were clear: Without the decision window, the team might have ignored the low-ranked alert or overreacted by disabling access during a critical onboarding week. The thirty-minute window allowed a calibrated response.

· · ·

Lena left the step-up challenge in place, issued a new multifactor enrollment flow, and revoked the exposed temporary password. She also added a just-in-time entitlement that only granted the Salesforce role after the security key was registered.

Three hours later, a second login attempt originated from a different foreign IP address and failed the new challenge. Someone had the old credentials. The incident was closed as a contained misfire rather than a false-positive storm or a real compromise.

The team updated the onboarding checklist to forbid password distribution by email and set the travel note field to required. The lesson was simple. Treat low-confidence alerts as hypotheses and place them on a clock: cadence, cross-verification, and small amounts of friction protect both trust and time.

The Quiet War

This war will not be televised. It will be personalized.

It will unfold in your feed, your workflow, your dashboards, and your thoughts. Its weapons are notifications. Its terrain is your attention span. Ambiguity is no longer just a strategic tactic. It has become a societal condition.

. . .

Institutions built to manage risk are being tested not by their adversaries' strength, but by their own susceptibility to confusion. The line between influence and interference has thinned to the point of disappearance.

To operate effectively in this domain, organizations need to become cognitive fortresses. Not walled off from information, but structured so they can interpret it without losing integrity. Resilience comes from process, not insulation.

That begins with culture. With leadership that values silence over noise, signal over scale, and truth over tempo. In this environment, the first casualty of confusion is confidence. Once confidence erodes, control follows. Ambiguity, left unchecked, not only clouds perception but also distorts it. It reshapes belief.

If ambiguity is the weapon, belief is the target. The next phase of this conflict will not be defined by what we see, but by what we trust.

In the next chapter, we will examine how belief is engineered, and how data ecosystems, narrative frameworks, and predictive algorithms now shape conviction as deliberately as markets shape price.

Chapter 3
The Machine's Blind Spots

The Confidence of the Unseeing

The machine sees everything but understands nothing. Its vision is immaculate. Its comprehension is absent. It recognizes patterns without meaning and produces confidence without understanding.

The dashboards glow with certainty. The probability scores rise. The anomalies are ranked and color-coded. The illusion looks perfect; a world neatly translated into metrics. But the machine's precision is not intelligence. It is repetition at scale. It mistakes correlation for causation, noise for nuance, and quantity for truth. We, in turn, mistake precision for understanding.

The result is a subtle but catastrophic error. Confidence divorced from comprehension. This is the machine's first and most dangerous blind spot. Not what it fails to see, but what it cannot know. Artificial intelligence does not hallucinate. It does not lie. It hallucinates because it lacks context. It cannot discern the weight of a word, the silence between gestures, or the meaning of an absence. Yet the smoother its outputs, the more we trust them.

The paradox of modern analysis is simple. The cleaner the data, the dirtier the assumptions.

The Illusion of Omniscience

Modern systems are built to see everything. Satellite constellations map every square meter of the planet. Algorithms analyze billions of transactions per second. Sensors capture every heartbeat on a factory floor and every click on a global network. But scale is not clarity. The more total our vision becomes, the less we seem to understand.

Vision without comprehension creates a dangerous illusion, the illusion of omniscience.

In security and intelligence, that illusion can be fatal. During the 2013 Boston Marathon bombing investigation, thousands of hours of surveillance footage were available within minutes of the attack. The abundance of visual data led investigators to believe that answers were close at hand. Yet it was precisely that volume, the multiplicity of plausible signals, that slowed early conclusions. Each frame offered a fragment of reality. There was no clear answer, meaning that human analysts had to reconstruct the story behind the movement. The cameras saw everything. Insight only arrived when someone understood what they were seeing.

Case Study: The 2008 Financial Crisis

In the years leading up to the collapse, financial institutions owned an extraordinary volume of risk models and algorithmic forecasts. Their dashboards glowed green. Every tranche, every derivative, every risk curve appeared precisely quantified. The machine's confidence was absolute.

What those systems lacked was context, the social and behavioral realities beneath the numbers. They saw rising property values as a trend, not a distortion. They saw high credit scores as a safety measure, not a simulation. When the models began to fail, the institutions that trusted precision over comprehension were the first to break.

The crisis was not a data failure. It was a failure of discernment. The numbers were accurate. The assumptions behind them were not. The machine had seen everything except the human behavior that made the numbers possible.

The Algorithmic Battlefield: Ukraine and Beyond

In modern conflict, algorithms now process intelligence faster than any analyst can. Satellite imagery, social sentiment, and intercepted signals are converted into probability matrices and risk heatmaps. On paper, this should sharpen decision-making. In practice, these systems often generate overconfidence.

Early in the 2022 invasion of Ukraine, both sides leaned on automated battlefield analytics. Predictive targeting systems highlighted likely troop movements and probable strike zones. Ukrainian forces learned to manipulate that vision. They understood the machine's habits. They used decoys, false heat signatures, and simulated radio chatter to lure algorithmic targeting away from real positions.

The machine, confident in its precision, was blind to deception. It saw patterns, not intentions. Human commanders who trusted those systems without critical oversight found themselves reacting to phantoms. This is the new fog of war, not the absence of data, but saturation without understanding.

The Corporate Mirror: Threat Intelligence and Over trust

Inside corporate security operations centers, AI-driven dashboards now aggregate thousands of threat indicators per hour. Each incident is assigned a severity score. Each alert is triaged through machine-learning filters and sorted before any human reviews it. The systems appear omniscient.

· · ·

When an algorithm elevates one event and suppresses another, few analysts question the reasoning. The authority of the interface, the sleek UI, and color-coded urgency create a familiar trap, automation bias. Analysts defer to the tool's confidence even when their intuition registers friction.

In a major breach investigation in 2021, a financial firm's threat-detection AI filtered out what appeared to be a low-confidence anomaly: an unfamiliar IP address communicating intermittently with a privileged server. The algorithm deprioritized it because it did not match known signatures, treating it as noise rather than risk. The result was a nine-month compromise that cost millions.

The system did not fail to see. It was unable to understand.

The New Discipline: Context as Countermeasure

To counter this illusion of mechanical confidence, organizations need to treat context as a core layer of intelligence rather than an afterthought. Every alert, every analytic, every model output should carry explicit lineage. Where it came from. What assumptions underlie it? What it cannot account for.

This requires rehumanizing analysis and training teams not to blindly trust dashboards, but to interrogate them. To ask:

What pattern is the machine showing me, and what is it missing? What human variable does this system not measure? What could this model not possibly know?

Machines can process signals. Only humans can interpret meaning.

The future of intelligence will not belong to those who see the most. It will belong to those who understand with the fewest distortions. The new advantage lies not in omniscience, but in epistemic humility, the nerve to doubt a perfect score, to question a clean dashboard, and to remember that clarity requires comprehension, not confidence.

The Mirage of Understanding

Machine cognition operates without consciousness, consequence, or curiosity. It models the world through probability, not purpose. To the machine, a child's laughter, a protest in the rain, and a market crash are all just clusters of data points with equivalent statistical weight. Meaning does not matter to the machine.

We have trained it to recognize what happens, not why. To detect the surface of behavior, not the substance of intention. In operational terms, this limitation appears as synthetic certainty, the quiet, persuasive tone of an algorithmic answer that feels objective but is not.

. . .

Executives trust it because it sounds neutral. Analysts cite it because it looks complete. Governments act on it because it feels inevitable.

Neutrality is not wisdom. A machine can model sentiment without feeling it, detect tension without sensing danger, and predict escalation without understanding stakes. This is not intelligence. It is instrumentation. Without interpretation, instrumentation becomes delusion.

Where Synthetic Certainty Misleads

Markets and models: high-frequency trading systems have produced sudden, severe price swings that their own models labeled as low-probability events. The systems were confident because the correlations looked stable. They missed fragile liquidity, human panic, and feedback loops that were invisible to the math.

Risk dashboards. Before major downturns, dashboards can glow green while underlying human behavior shifts. Models see positive momentum. They do not see fear, pride, or denial inside boardrooms and households.

Operational lesson: A precise probability can mask a brittle assumption. Add human scenario checks that ask what the model cannot know.

The Self-Reinforcing Bias

Modern systems promise prediction. Often, they deliver repetition.

Predictive policing offers a clear example of how this happens, and why synthetic certainty is so dangerous.

When the Los Angeles Police Department adopted a program known as PredPol in 2011, the idea was straightforward. Use data to forecast where crime is most likely to occur. To do this, the system analyzed years of police reports, searching for patterns of time, location, and offense type. It divided the city into thousands of small digital squares and assigned each a probability score. The highest-ranked zones became hot spots that guided daily patrols.

On paper, it looked like progress, a data-driven compass that could direct limited resources efficiently and objectively. But in practice, the model did not study crime. It studied policing.

It learned from the data it was given, and the data reflected where officers had previously been, not where crime had necessarily occurred. Neighborhoods with higher police presence generated more reports, arrests, and apparent evidence of risk. That new data fed back into the system, which confidently declared those same neighborhoods even riskier.

. . .

The algorithm's precision became a feedback loop. Each patrol validated the model, not reality. Even minor citations became data points confirming the model's prediction. Confidence scores climbed. Community trust fell.

What began as a neutral tool evolved into a mirror, reflecting institutional patterns with mathematical authority. It amplified the very biases it was meant to remove. Areas with fewer police officers produced fewer reports. The model learned to treat them as low risk even when victim surveys told a different story. The machine's lens, trained on past enforcement, could not see what was invisible to its data, crimes unreported, tensions unseen, distrust unmeasured.

By 2019, independent audits by the Los Angeles Times and the Human Rights Data Analysis Group showed that PredPol's accuracy was illusory. Crime rates had not improved. Patrol patterns had hardened along preexisting social lines. In a 2019 report, the LAPD's own Inspector General concluded that predictive policing had no demonstrable effect on crime reduction and had created a perception of targeted over-enforcement. By 2020, the department ended the program.

Los Angeles was not alone. Chicago's Strategic Subject List sought to identify people most likely to commit or become victims of gun violence. The model drew on arrest histories and social networks, assigning risk scores that appeared empirical but were based on incomplete context. Once labeled, residents found themselves in a kind of algorithmic limbo, surveilled more heavily because the model deemed them high risk and scored them accordingly. A 2019 RAND Corporation evaluation found no evidence that the system reduced shootings or improved safety. It did deepen mistrust between police and the communities they served.

In Oakland, a team from the Human Rights Data Analysis Group simulated PredPol's algorithm using historical data. Their findings were stark. The system would have sent officers back to the same neighborhoods repeatedly while ignoring others with comparable crime levels but less enforcement history. The algorithm was not biased by intent. It was obedient to its data. It did not know it was amplifying inequality. It simply replicated what it had learned.

This is what happens when precision is mistaken for perception. The models produced impressive maps and dashboards glowing with confidence. The confidence was synthetic. The machine could not question its own inputs or recognize that its predictions fed on its own outputs. The circle closed, and the system congratulated itself on its accuracy.

The greater danger was not statistical error, but epistemic illusion, the belief that a number generated by a machine is more objective than a judgment made by a person. Executives trusted it because it sounded neutral. Analysts cited it because it looked complete. Governments acted on it because it felt inevitable.

The map is not the world.

Predictive policing became a symbol of a larger truth about algorithmic decision-making. Machines cannot see intent. They cannot weigh trust or fear. They can measure what happens. They cannot measure what matters. Without context, their predictions become recursive, repeating past actions under the banner of foresight.

The same pattern now threatens to repeat in cybersecurity, financial modeling, medical triage, and threat intelligence. Systems trained on biased or incomplete data can reproduce errors with perfect fidelity, presenting old patterns as new insight. The outputs are confident and persuasive. The bias hides behind probability scores and color-coded dashboards.

From a security perspective, this is not only a technical problem. It is a cognitive one. A machine that cannot question its assumptions will always confirm them. A human who cannot ask the machine will eventually believe it.

The lesson of predictive policing is not only about data ethics. It is about epistemic discipline. The future of intelligence depends less on how much the machine sees and more on how rigorously we interpret what it shows us. When the system insists it knows where to look, the wiser course is to look there and somewhere else as well.

Chapter 4

Open-Source Intelligence and Social Media

Sentiment Without Meaning

The promise of open-source intelligence, OSINT, is unprecedented visibility. Millions of posts and interactions form a real-time pulse of global behavior. The same scale that makes this ecosystem valuable also makes it volatile.

When algorithms measure sentiment by velocity, not veracity, they produce something like sentiment without meaning, a map of emotion unmoored from reality.

Corporate security teams have learned this the hard way. During protests in 2022 after a high-profile corporate layoff in Southeast Asia, several multinationals relied on automated social listening tools. They were monitoring potential threats near their facilities. One firm's dashboards registered a sharp spike in negative sentiment. The algorithm translated that spike into a high threat probability and escalated an alert to leadership.

Executives, fearing unrest, ordered a lockdown of the local office. The response was swift and unnecessary. A subsequent manual review found that most of the posts originated from newly created accounts linked to an activist group that was reposting identical messages in multiple languages. The campaign was coordinated, not organic. The anger was algorithmic theater.

The lesson was blunt. Velocity is not veracity. Machines rank the speed of information spread. Commanders need to understand the origin of the information itself.

Algorithms detect language and motion, not intent. A surge of emotional language, anger, sadness, or fear can look identical in data form. The underlying reality may be synthetic. In this ecosystem, misinterpretation at machine speed can trigger costly overreaction, false alerts, unnecessary lockdowns, or reputational harm when companies respond publicly to events that never existed.

The remedy is operational discipline. Every automated alert should pass through a verification gate, a brief but structured human pause where analysts check provenance, countersignals, and the integrity of sources. In OSINT, the goal is not to respond faster. It is slower cognition. The first one to pause is often the first one to see clearly.

Computer Vision at the Edge

Vision systems now patrol campuses, logistics hubs, and data centers. They detect motion, classify objects, and alert operators within seconds. The favored term is situational awareness at the edge. High accuracy does not equal high understanding.

A well-known incident at a global logistics firm illustrates this. An AI perimeter system detected an unusual assembly outside a European distribution center. The classifier, trained on movement patterns, labeled the event a potential protest. Security protocols escalated the alert to corporate headquarters. Local law enforcement was contacted in advance.

When responders arrived, they found not a protest but a community memorial for a former employee who had died in an accident the week before. Candles and standing figures mimicked the crowd density the model had been trained to flag. The system had seen correctly and understood nothing.

This is recognition without comprehension. Computer vision excels at repetition. It can classify thousands of images per minute, track individuals, or identify loitering. The moment an edge case appears, something outside its training data, the precision collapses.

At a refinery, an automated camera network once flagged a person down near a restricted area. The alert triggered an immediate emergency response. Upon arrival, teams found a thermal blanket and a sandbag. The low-light classifier misclassified the outline of a human body, and the algorithm's confidence score still exceeded 95% — resulting in a false alarm that cost time, coordination, and morale.

The flaw was not in the math. It is assumed that vision equals understanding.

Corporate environments rely on machine vision for safety and security. The real risk lies in uncritical trust. A camera can detect presence, not purpose. Motion, not meaning. Without human interpretation, it cannot distinguish between a vigil and a protest, or between an object and a person.

Pair each sensor with its interpretation rules. Context must travel with the data. Require human review before escalating any response that involves human interaction, law enforcement notification, or reputational stakes. The first responder's most valuable tool is not the alert. It is discernment.

Strategic Warning and the Illusion of Foresight

In a hyperconnected risk environment, every organization now operates an early-warning system—cyber, physical, and geopolitical data streams into unified dashboards. The intention is to catch anomalies before they become crises. The effect is often an early warning overload.

Color-coded urgency creates the illusion of comprehension. A heatmap of potential threats appears scientific and rational. Behind those tiles often sits a tangle of unverified inputs, social media rumors, news aggregators, sensor noise, and third-party alerts.

During global supply chain disruptions in 2021, several Fortune 500 companies learned this the hard way. Their risk intelligence platforms began flagging red-level anomalies in regions where protests and cyber incidents appeared to coincide. Executives interpreted the overlap as coordinated targeting and enacted costly continuity plans. They rerouted shipments, delayed operations, and increased physical security.

Weeks later, analysts found that the anomalies shared timing, not cause. The cyber events were unrelated phishing campaigns. The protests were spontaneous and not linked to the companies involved. The pattern existed only inside the algorithm's compression of reality.

· · ·

The same year, a disinformation collective used open networks to simulate brand-targeted activism against multiple corporations. Dozens of coordinated accounts posted threats of protest using authentic logos and staged event details. The content triggered automated alerts across corporate intelligence systems. Executives responded with statements and advisories. That overreaction was the adversary's fundamental objective.

This is weaponized ambiguity, the deliberate flooding of open systems with plausible but misleading signals to exploit algorithmic pattern recognition. The model sees a pattern. The adversary sees your reaction.

Corporate intelligence units now face a tempo dilemma. Move too slowly and risk genuine loss. Move too quickly and become part of the deception.

Operational discipline is the only stable counterweight. Integrate deception detection into tradecraft. For every anomaly flagged by automation, ask a structured question. How could an opponent make this model wrong on purpose? That single habit reintroduces doubt into a system optimized for confidence, returning cognition to a process.

Across these examples, whether social sentiment, computer vision, or strategic warning, the pattern repeats. Machines amplify the signal. Humans must restore meaning. The challenge is not blindness. It is clarity distortion, a world seen in high definition and interpreted in low context.

Accurate intelligence, corporate or national, depends less on seeing faster and more on seeing with restraint. Every system can be fooled by its own precision. Every confident dashboard must answer a simple question: "What, exactly, am I looking at, and who wants me to see it this way?"

How to Resist the Mirage

The danger of machine precision is not only error. It is persuasion. A number, a score, or a color-coded alert carries the quiet authority of objectivity. Data without provenance or automation without context can turn precision into illusion.

To resist that mirage, organizations need to embed discipline into the architecture of their analytic systems. These are not abstractions but operational countermeasures against synthetic certainty.

Provenance on Every Output

Every algorithmic score, label, or recommendation should carry its lineage. The data source. How it was derived. The transformations applied. The assumptions are baked into it. The blind spots known to affect it. Without traceability, confidence is unearned.

. . .

In intelligence work, provenance is the foundation of trust. A report without source integrity is gossip with a header. Machine outputs are no different. Yet they often arrive stripped of history, detached from the scaffolding that produced it.

When a dashboard reports a threat level of 83%, the first question is not what that number means. It is where it came from. If you cannot trace it, you cannot trust it.

Every machine output needs its birth certificate. Provenance turns raw data into accountable intelligence. It forces transparency back into systems built for speed and gives analysts a way to question not just the conclusion, but the process that produced it.

The Two Key Turns for Action

In nuclear command, two officers must turn keys at the same time to authorize a launch. The system introduces redundancy and restraint, a safeguard against single-point failure under pressure.

The same logic should apply to algorithmic decisions. Any high-consequence action, a facility lockdown, a large financial trade, or a crisis-stage public statement should require agreement between a machine output and an independent human hypothesis. If they align, proceed. If they diverge, pause and investigate the gap.

. . .

This discipline reintroduces friction into systems built for immediacy. It turns speed from a reflex into a privilege, something earned through verification rather than assumed by design.

In corporate security, this approach has prevented false alarms, unwarranted evacuations, and reputational damage. Automation should accelerate execution, not judgment. The second key ensures that the first is turned intentionally.

Context Loops in Automation

The strongest machine learning models are not those that learn fastest. They are the ones who learn accurately, especially from their failures. Most organizations track success metrics and ignore the cases where the system was confidently wrong.

Context loops close that gap. They feed verified outcomes back into the model and weigh false confidence more heavily than correct predictions. Every mistake becomes a teacher. Every near miss becomes data for humility.

This feedback culture turns automation from oracle into a learner. In one global security operations center, after-action reviews now include an error brief alongside success summaries, detailing where the machine misread sentiment, misclassified imagery, or overpredicted risk. These cases are not punished. They are studied.

Confidence without correction is a liability. Context loops are how organizations evolve discernment at scale.

Red Team the Model

Every defensive system needs an internal adversary. Red teaming an algorithm is not an act of distrust. It is an act of stewardship.

A standing analytical cell should deliberately break the model, probe it with adversarial data, test edge cases, and explore how small inputs can drive significant failures. This is not abstract gaming. It is a practical inoculation.

When a cybersecurity firm began testing its AI threat-detection platform using deliberately misleading data, the results were sobering. Minor input changes, misspellings, alternative date formats, and mirrored IP strings could hide entire campaigns from detection. Those findings drove redesigns that made the model more resilient and the team more skeptical.

The output of a red team is technical and cultural. Publish the findings internally as a How to Fool Us memo. Confidence should be tested like armor, not admired, but struck.

Clarity Audits

Accuracy alone is not enough. Analysts also need to understand why a system concluded. Clarity audits are structured reviews that treat interpretability as seriously as precision.

Every quarter, teams can select a sample of automated outputs, alerts, rankings, and forecasts, and require analysts to explain the rationale for each. Can they explain why the top three alerts were ranked above the next three? If not, trust in that system should be downgraded until transparency improves.

These audits build a habit of reflection in environments dominated by velocity. They surface hidden dependencies, expose model drift, and remind leadership that black box systems are not partners. They are dependencies. If an analyst cannot explain a machine's reasoning, the organization has lost control of its own cognition.

Interpretability is not a luxury. It is governance.

Decision Latency by Design

In a crisis, speed feels like control. Often, haste is just anxiety in motion.

. . .

The final discipline is to engineer decision latency, a deliberate pause for verification in any system that handles high-risk decisions. Not every action needs to be instant. A ten-minute delay before a mass notification. A secondary confirmation before a public statement. These micro-hesitations can prevent macro failures.

Corporate security environments are saturated with dashboards that measure response time down to the second. Speed without context can amplify risk rather than reduce it. Decision latency turns time into a tool of judgment rather than a performance metric. Speed is not strategy. It is a setting. In the age of algorithmic acceleration, the ability to wait may be the last real form of control.

Together, these disciplines form a framework of operational skepticism, a culture that values comprehension over convenience. In a world of confident machines, the organizations that endure will not be the ones that see fastest. Instead, it'll be the ones who verify what they see. The future of intelligence will not rest on superior data, but on superior doubt.

The Machine and the Mirror

Machines excel at scale and recall. They can read every log file, cross-reference every transaction, and monitor every camera feed without fatigue or distraction. They never forget. They never rest. Their strength lies where humans are weakest: the ability to surface weak signals across vast domains and connect subtle patterns that no individual could hold in mind.

In that role, they are indispensable filters, amplifiers of possibility. They must never be confused with arbiters of truth.

Machines are fluent in correlation and blind to consequence. Their efficiency ends where meaning begins. The most dangerous error in the age of automation is not technical. It is a psychological mistake, mistaking computational reach for comprehension.

Humans are not exempt from blindness. Intuition is powerful and fallible. Narrative is persuasive and selective. A confident analyst can be as dangerous as an overconfident model. Cognitive biases such as confirmation, anchoring, and recency bend perception long before any machine processes the data.

The real frontier is not human against machine. It is bias against bias, aligning mechanical pattern recognition with human contextual judgment so each helps correct the other's distortion.

That partnership must be intentional. Machines counter human myopia by revealing scale. Humans counter machine blindness by restoring meaning.

. . .

Not every error warrants the same level of scrutiny. In low-stakes, repetitive environments, automation's speed is an acceptable trade for imperfection. If a recommendation engine mislabels a photo or misroutes a noncritical alert, the damage is minimal. Apply the same margin of error to layoffs, supply chain shutdowns, or active threat response, and it becomes existential. The work is calibration. Trust must scale with consequence.

Some advanced models, such as intensive learning systems, cannot fully explain their internal reasoning. In those cases, organizations need external guardrails, conservative action thresholds, diverse training data, and mandatory human review for decisions with ethical or operational weight. The absence of full explainability must never be mistaken for trust.

Lessons From the Field

In one security operations center, an AI platform flagged a low-confidence anomaly from an unusual IP address. The dashboard ranked it near the bottom of a long list of alerts. On another day, it might have been ignored.

An analyst remembered that a new contractor had been onboarded the previous week. She checked the identity pipeline and confirmed that the IP matched the contractor's recent travel, and that a newly created account was missing metadata due to a permissions lag. The event was quietly re-ranked as high risk and blocked.

. . .

That one act of human memory prevented what could have become a long-term credential compromise. The algorithm was not wrong. It was incomplete. The gap between accuracy and understanding was a person with context.

In another case, a facility command post relied on automated sentiment tools that flagged a sudden surge in negative posts directed at a local office. The system escalated posture to heightened alert. Local leaders were prepared to activate protest protocols.

Before proceeding, a social media analyst looked at the source accounts. Most were less than a week old and posted identical language and images recycled from unrelated events. It was a botnet simulation, not a protest.

What could have become a costly physical security mobilization turned into a digital counter-messaging and law enforcement coordination effort. The key move was simple. Someone asked who was speaking, not just what was said. Sentiment is not intent. Algorithms cannot tell the difference.

In a third scenario, a corporate crisis cell reacted to a viral video clip that appeared to show chaos outside one of its regional offices. The classifier labeled the scene a riot and triggered an executive alert. A watch officer decided to pause escalation. She geolocated the footage, checked the metadata, and confirmed the clip was two years old and from another continent.

Ten minutes of skepticism prevented ten days of reputational damage.

These stories share one thread. The major missteps were avoided not by technology, but by refusing to surrender judgment to it.

The Standard to Adopt

Every machine output should be treated as a hypothesis, not a verdict. Probability is not proof. Scores, labels, and alerts must include context, data lineage, confidence ranges, and known limitations to ensure interpretation remains possible.

Trust must match consequence. A small automation error in payroll reconciliation may be annoying and fixable. The same mistake in a threat forecast can spark public panic. The higher the stakes, the higher the bar for verification.

Interpretation should be collective. High-performing intelligence teams often rotate the role of designated skeptic, someone tasked not with adding answers but with challenging assumptions. That simple ritual institutionalizes dissent and keeps confidence from hardening into dogma.

· · ·

Above all, organizations need to normalize intellectual humility. We do not yet know whether it should be a valid operational state, rather than a mark of failure. Acknowledged uncertainty is a form of strength. It signals awareness of the limits of knowledge, which is the first condition for learning.

The machine can help us see. Only humans can help us understand. The path forward is not to abandon automation, but to surround it with discipline, context, and courage.

The Human Countermeasure

Human context remains the only reliable antidote to mechanical blindness. Machines model behavior, and humans interpret intent. Intent lives in the spaces between data points, in hesitation, silence, gesture, and choice. Those gaps are invisible to code and legible to empathy.

Empathy here is not sentimentality. It is contextual literacy, the ability to read meaning through people, not around them. A seasoned analyst knows when a statement hides fear, when a pattern hints at fatigue, when a spike in metrics signals grief rather than threat. Empathy lets a leader interpret tone on a crisis call or recognize that a quiet employee's withdrawal carries more weight than any dashboard.

In that sense, empathy is an analytic instrument in its own right. Data can tell you that something happened. Empathy can say to you why it mattered.

. . .

In corporate and geopolitical intelligence, that distinction separates foresight from failure. During the early days of COVID-19, several logistics firms misread employee hesitation as absenteeism instead of fear. Those who listened kept supply lines intact by addressing emotional truth, not just operational gaps.

The reconciliation between computation and compassion will shape the future of intelligence work. Without it, we risk building architectures that know everything, yet understand nothing.

The Error of Perfect Data

True clarity is not the absence of error. It is the capacity to learn from it. Perfect data is a myth... and a dangerous one. Perfection smooths over the texture that gives intelligence depth. Ambiguity, contradiction, and uncertainty are not flaws in analysis. They are its raw material.

When we force the world into algorithmic neatness, we erase the irregularities that enable insight. A sensor that never errs is a sensor that never learns. A model that never doubts is a model that never questions. Understanding grows in imperfection, in the pause between data points where interpretation starts.

. . .

Human insight is born in error, in the friction between expectation and reality, between what the machine predicts and what occurs. To understand imperfection is to understand reality. Precision alone is sterile. Understanding is alive.

Seeing With Intent

To see like a machine is to measure. To see like a human means to understand.

The challenge of our time is not teaching machines to think. It is remembering how to think for ourselves. In an age dominated by automation, our last real advantage is intentional attention, the discipline of seeing with purpose rather than just with precision.

The analyst who resists the algorithm's certainty, the leader who asks why before how fast, the system designer who measures success by veracity rather than volume, these are the custodians of clarity. Clarity, at its core, is moral. It is not about knowing more. It is about knowing what matters.

That distinction marks the boundary between intelligence and imitation, between awareness and automation, between meaning and math.

The machine's blind spots are not its failure. It is their nature. Our failure would be to inherit them as our own.

. . .

If the previous chapter is about blindness, this one points toward design. The next step is structural. How do we build systems that preserve meaning instead of eroding it? How do we architect clarity, culturally and technically, so that human insight is amplified, not replaced?

That is where we turn next, to the architecture of clarity and the deliberate design of systems that see not only what is visible but also what is true.

Chapter 5
Tradecraft Reborn

Every system, no matter how advanced, begins and ends with a human being. Automation can accelerate detection, flag anomalies, and generate probabilities, but the loop always closes on a person who must decide what to believe and what to do. The machine can illuminate. Only humans can interpret. The operator remains accountable, even when the system is autonomous.

Across history, intelligence work has been a human discipline disguised by its tools. Tradecraft was never really about encryption keys, miniature cameras, or encrypted radios. It was about perception, the ability to notice what others missed and understand what others misread.

During the Second World War, British analysts at Bletchley Park used machines to decode German transmissions, but it was human intuition that found patterns in Enigma's operational errors. The systems cracked messages. People cracked the meaning.

In our time, the pendulum has swung toward total automation. We trained machines to notice everything and, in the process, untrained ourselves to see anything. Sensors became our eyes. Dashboards became our instincts. Operators became caretakers of systems rather than interpreters of events. Telemetry replaced tradecraft.

In a corporate security center, analysts may monitor a wall of screens that display summaries of access logs, camera feeds, license plate detections, and more. Their role often narrows to acknowledging alerts rather than discerning patterns. When a breach occurs, the postmortem usually shows that the indicators were present all along, buried in the noise, invisible not because they were hidden, but because no one was truly looking.

No system can remove the need for human discernment. Each alert, analytic, and model prediction still hinges on one irreducible act. Someone must decide what it means. At that moment, the loop begins again, not with more data, but with renewed attention.

From Awareness to Discernment

Awareness is knowing what is happening. Discernment is

understanding why it matters. Awareness detects motion. Discernment extracts meaning.

The world has mastered awareness. In corporate risk, public safety, and national defense, sensors and platforms detect even minor behavioral fluctuations. What we have lost is discernment, the capacity to understand the significance of what we see.

The 2008 financial crisis made this painfully clear. Analysts across the industry had visibility into mortgage default rates, collateralized debt patterns, and leveraged exposures. The awareness was total. Very few grasped the systemic connections that made the collapse almost inevitable. The data were visible. The meaning was opaque.

Discernment is not a technical function. It is a human act of synthesis that connects behavior, motive, emotion, and context into a coherent picture. When an insider threat alert flags an unusual download, discernment asks why now, why this file, why this person. When an anomaly detection algorithm flags traffic from a foreign IP, the detection process considers not only the threat model but also the corporate context, a vendor's remote access, a legitimate software update, or a false positive triggered by travel.

Discernment cannot be automated because it involves perspective. It must be cultivated through mentorship, repetition, and deliberate exposure to ambiguity. The modern analyst needs to be bilingual, fluent in both human nuance and machine logic. This hybrid practitioner treats algorithms as instruments, not oracles. They translate statistical correlation into behavioral inference. They move between empathy and evidence, between code and context.

The future belongs to those who can write Python and read people.

The Convergence Principle

Reborn tradecraft is not nostalgia. It is convergence, the fusion of human discipline and machine augmentation.

The core traits that defined great intelligence work, patience, pattern recognition, discretion, and intuition, remain as vital now as in the Cold War. The arena has changed. The alleyway has become the algorithm. The dead drop is now a data cache. The brush pass has turned into a metadata exchange.

During the Cold War, CIA officers mapped human networks to understand how influence flowed through communities. Today, practitioners map digital ecosystems in which influence spreads through hashtags, retweets, and algorithmic amplification. The task is the same. Follow the signal. Distinguish the authentic from the orchestrated. Find the intent hidden inside the pattern.

. . .

This convergence dissolves the boundary between intelligence and technology. The question is no longer whether human judgment or machine analysis is superior, but how they can work together to produce clarity. The hybrid analyst embodies this reconciliation. They use automation for reach and scale, tempering it with restraint and moral judgment. The goal is not automation of insight, but augmentation of awareness.

Reclaiming the Craft

To reclaim tradecraft is to recover discipline in perception. Gadgets did not define the masters of old. Habits defined them. They looked longer than others cared to. They listened for what was not said. They withheld judgment until they had earned it.

These are not lost arts. They are neglected ones. In an age that prizes speed over understanding, patience becomes a competitive advantage. Systems should be designed to reward careful thought rather than instant reaction. Analysts should be trained to treat ambiguity as a signal, not as a flaw in the model.

In cyber defense, a partial pattern, a subtle time variance in network traffic, a repeated failed login at dawn, can hold more truth than a dozen clear-cut alerts. Only the patient analyst sees it.

· · ·

Reclaiming the craft is not a form of resistance to progress. It is alignment, teaching technology to match the rhythm of human curiosity. We should adapt machines to the pace of thought, not force thought to match the pace of machines. The aim is not only to see more of the world, but to think more wisely within it.

The Checklist that Cut Twice

Operating room four ran five minutes behind schedule—a routine laparoscopic cholecystectomy. The team had worked together hundreds of times. The attending surgeon preferred a tight tempo. The anesthesiologist signaled stable vitals. The nurse circulator announced the first count complete. Everyone knew the script.

Signal confusion. After the gallbladder was freed and bagged, the surgeon asked for the closing suture. The scrub tech hesitated. The whiteboard instrument tally was off by one radiopaque sponge. The circulator suggested the missing count might be due to a recording error during an early handoff. Imaging would add 15 minutes and disrupt the case's rhythm. The first-pass checklist had been completed. The patient was stable. The story could be clerical noise.

The charge nurse entered the room and invoked a second verification ladder that the unit had adopted following a recent near-miss. It called for a short time box. Pause for two minutes to re-run the count with a fresh pair of eyes. If still off, perform a rapid field sweep. If still off, order imaging before closure.

The two minutes felt long. The second count was still short by one. The sweep of drapes and floor found nothing. Imaging was ordered. The surgeon agreed, irritation visible. The X-ray appeared on the wall monitor. A faint marker sat in the right upper quadrant, blurred by insufflation gas. The sponge had slipped behind a retracted fold. Retrieval took less than a minute.

Outcome and lesson: No harm was caused to the patient. The team closed on time. The hot wash was brief. The original count had been accurate. The sponge had migrated during suction. One slight deviation in protocol would have set the patient up for a second, riskier procedure days later.

The lesson held. Rituals are not theater. They are engineered to catch low-probability, high-consequence failures. The second cut checklist did not add complexity. It added a deliberate pause with a rising ladder of verification that ended in imaging when uncertainty remained.

The team adjusted the whiteboard process to show both counts and the time of last reconciliation. The attending apologized to the room for the flash of impatience and praised the nurse for pulling the cord. Composure is a team sport. The checklist worked because the culture allowed anyone to stop the line.

The Return of the Human Edge

In an age when machines can predict behavior and autonomously generate intelligence products, the last real frontier of unpredictability remains the human mind. Authentic tradecraft lives there. It is the capacity to empathize with an adversary, perceive a pattern hidden in noise, and act with both logic and moral weight.

A familiar scene from fiction captures this in concentrated form.

During the diner scene in *The Bourne Identity* (2002), Jason Bourne calmly inventories his environment. He knows the license plates of the cars outside, gauges the bouncer's build, registers that the waitress is left-handed, notes the likely location of a weapon, and reads his own physical limits. He is not improvising random trivia. He fuses perception, pattern memory, and action into a single loop.

That is not machine prediction. It is human composure and real-time meaning-making.

It matters for three reasons.

· · ·

First, symbiosis, not substitution. Sensors and models can surface possibilities. Only a human can weigh context, intent, and consequence in the moment.

Second, trust is earned in the last meter. When uncertainty spikes, teams default to the person who can observe clearly and make decisions with moral weight.

Third, the loop can be trained. Drills that require rapid observation, inference, and action embed this "Bourne loop" in muscle memory.

In an age of autonomous analytics, the final competitive edge is human tradecraft, the ability to empathize, see patterns in noise, and act with judgment. Machines can narrow options. Only people can choose what ought to be done.

No algorithm can replicate courage or restraint. When a corporate security director decides whether to evacuate a site amid conflicting data, or when an intelligence officer decides whether to reveal a source, the decision is not purely computational. It is human. These moments test judgment, not processing power.

The machine will always be faster. The human can still be wiser. Tradecraft reborn is not a rejection of technology. It is the restoration of humanity to its rightful place inside it. The loop may begin with data. It still ends, as it always has, with a decision and a person who bears its consequence.

Chapter 6
Behavioral Signatures

Every human leaves a trace. Even silence has cadence. Every action, hesitation, and deviation contributes to a language of behavior that reveals more than words alone. Behavior writes its own script in timing, tone, and rhythm. It surfaces in how a person responds to stress, repeats a task, or unconsciously alters phrasing in an email.

These are behavioral signatures, the fingerprints of attention. They are not about who a person is, but how a person is being.

Behavioral signatures underpin modern intelligence. They reveal the tempo of intent beneath surface activity. In the digital world, a delayed response, a sudden change in routine logins, or an abrupt shift in communication tone can speak volumes. To the untrained eye, these appear as ordinary fluctuations, noise in an endless stream of activity. To those who practice disciplined observation, some signals reveal motive, state, and direction.

Consider insider threat detection in a corporate environment. A trusted engineer begins working odd hours, pulling repositories unrelated to current projects, and using a private device near company systems. Each behavior in isolation is trivial. Together they form a pattern. What matters is not only what the individual does, but when and how those actions change.

Behavioral intelligence focuses on changes in rhythm, the syncopations that hint at intent. In earlier eras, investigators followed footprints in the snow. Today, they analyze frequency patterns in data streams.

The Language of Emotion

Emotion moves through digital systems like heat through air. A single burst of outrage or empathy can cascade across a network before reason catches up. Social platforms, stock markets, and internal corporate chats all run on emotional dynamics. Information moves fastest where feeling is strongest. To understand behavior in these environments, you need to sense emotional temperature.

The best analysts are the ones who can feel before they can count. They detect tonal shifts in language, subtle rises in community tension, or quiet withdrawal, long before metrics confirm that anything has changed.

During the early stages of the Arab Spring, it was not hashtags or trending charts that signaled upheaval. It was the shift in emotional resonance. Posts moved from frustration to resolution, from grievance to coordination. Machines could map the words. Only humans could feel the escalation.

Algorithmic sentiment analysis can measure positivity or negativity in text. It cannot capture voltage, the emotional force that drives behavior. It counts vocabulary, not vibration. Proper understanding requires sensing the charge behind words.

Emotion is not noise in the system. It is the heartbeat. Analysts who attune to its rhythm can sense change before any statistical model predicts it. In financial markets and online discourse alike, the emotional pulse often precedes the measurable trend.

Anomalies as Stories

In many systems, anomalies are treated as errors, outliers to be corrected or filtered. In behavioral intelligence, anomalies are places where truth often lives. The average conceals. The anomaly reveals. Patterns describe structure. Anomalies tell stories.

. . .

A stark example emerged in aviation security before September 11, 2001. Several flight school instructors reported that foreign students wanted to learn to fly planes but showed no interest in learning to land. Each case, viewed in isolation, appeared eccentric but not alarming. The pattern only became clear later, when it was too late to act. The anomaly had been visible all along. Routine muted the signal.

The same principle applies in organizational risk. A team member who suddenly withdraws from meetings, stops collaborating, and avoids communication is not just an HR concern. They may be an early sign of burnout, disillusionment, or malicious intent. Each behavioral break carries a narrative. The analyst's task is not to smooth it away, but to interpret its meaning.

A system without anomalies is a system without humanity. Human behavior naturally oscillates. Each fluctuation, pause, surge, or deviation is a form of communication. What appears as a data point is often a confession in disguise.

The Pulse of Attention

Attention leaves its own residue. The way people move through information, where they linger, what they skip, whether they pause or rush, offers a map of their cognitive landscape. These micro patterns form attention signatures, the traces of thought translated into motion.

· · ·

Across populations, attention patterns become collective rhythms. A workforce that once read internal updates carefully may begin to skim them. A community that once engaged in thoughtful debate may shift to reacting only to headlines. A marketplace that once absorbed long-term analysis may suddenly trade on rumor.

These tempo changes reveal underlying shifts in trust, fatigue, or morale.

In digital ecosystems, attention is often the first indicator of fracture. Before a cybersecurity breach, internal attention tends to splinter. Alerts go unread. Procedures are skipped. Vigilance fades. Before a social movement emerges, attention harmonizes. Language aligns. Focus narrows. Discourse coheres around shared symbols.

Attention functions as both a mirror and a map. It reflects the current state and hints at what comes next.

Seeing Through the Human Pattern

Seeing behavior clearly requires more than data collection. It uses empathy as an analytical tool. Empathy allows an observer to feel motion within behavior, to recognize when a person shifts from curiosity to concern, from cooperation to resistance. It turns data into understanding.

. . .

Behavioral signatures are not static identifiers. They are living expressions of the state. They remind us that every metric has a heartbeat. A spike in productivity may signal focus. It may also indicate desperation. A sudden drop in engagement may suggest distraction or disillusionment. Without empathy, both look identical.

This is what separates human-centered intelligence from algorithmic analysis—machines record movement. Humans perceive meaning in motion.

To understand behavior is to stand within it, not above it. It cannot be done solely from the tower of analytics. It requires immersion in human experience, where context, motive, and feeling converge. The analyst has to learn to inhabit the pattern, not just observe it.

The Analyst as Sensor

The modern analyst is not just a data processor. They are a sensor of significance. Their job is to interpret subtle shifts before they harden into visible trends. That work demands more than technical skill. It demands embodied intuition, an awareness sharpened through repetition, reflection, and mentorship.

Technology extends its reach. It does not replace their perception. Perception still must be earned through the slow accumulation of context.

. . .

The analyst listens for the tremors beneath the metrics-a tone shift in an email, a change in the rhythm of team communication, the subtle emotional undertow in a conversation thread. These are the vibrations of human systems, barely perceptible and rich with meaning.

In one case, a counterintelligence officer detected espionage not through a data breach, but through a conversation. The suspect's language became overly careful. Their usual humor disappeared. Their phrasing turned stiff and bureaucratic. The change was microscopic and unmistakable. The machines ignored it. A trained human did not.

This is the modern art of observation, translating feeling into foresight and foresight into form. In this sense, tradecraft becomes a discipline of listening. Before behavior becomes history, it begins as a signal.

The analyst's task is to hear it clearly and patiently, before the world catches up.

Chapter 7

Cognitive Countermeasures

The First Casualty

The first casualty of data overload is composure. Before logic falters and before analysis collapses, calm is the first thing to disappear. When the mind is flooded with signals, equilibrium fails first. Under pressure, attention contracts, time perception distorts, and emotion rushes in to fill the space where certainty once sat.

What we often label as bias is not always intellectual failure. It is the brain's defensive response to speed. Overload is not only a technical problem. It is physiological. The human nervous system was never designed to engage continuously with high-speed information environments.

When an analyst or leader faces dozens of alerts per minute, heart rate rises. The prefrontal cortex—the center of planning and judgment—yields ground to the limbic system, which governs emotion and the fight-or-flight response. Decision quality suffers not because the individual lacks skill, but because the brain is shielding itself from excess.

In 1983, Soviet Lieutenant Colonel Stanislav Petrov received a computer alert showing multiple U.S. nuclear missiles inbound. The protocol told him to retaliate. His own senses told him something was off. He paused. He reasoned that a real strike would be larger. His composure prevented catastrophe. The alerts were false, triggered by a malfunctioning satellite sensor. That night remains one of the clearest examples of cognitive countermeasures at work: clarity preserved by restraint.

Cognitive countermeasures protect thought itself. Every analyst and leader now operates in an attention battle space where the real opponent is often not misinformation but misperception.

Bias as Efficiency

Bias is not error. It is efficiency. It is the brain's compression algorithm—its shortcut for navigating complexity with limited time. Bias allows decisions under partial information. It is a survival adaptation. What protects us in one context distorts us in another.

. . .

Confirmation bias, authority bias, and availability bias all stem from the same cognitive bias. The 2003 invasion of Iraq is the familiar example. Analysts under political and temporal pressure over-weighted preexisting assumptions about weapons programs. The analytic machinery worked. The cognitive calibration did not. Bias was not the absence of intelligence. It was intelligence misdirected.

The disciplined practitioner manages bias the way a pilot manages wind, not by denying it, but by compensating for it. Every mind has cognitive drift. The key is knowing which way it pulls.

In cybersecurity investigations, analysts must counteract confirmation bias by deliberately seeking evidence that disconfirms their findings. When they see suspicious traffic, the instinct is to match it to known attack patterns. The countermeasure is structured doubt—reframing the data through alternative frameworks to test whether the theory still holds.

Reflection is calibration. Bias cannot be erased, but it can be managed through awareness, rotation, and humility. Skilled analysts do not trust their first conclusion. They test its edges until they either break or strengthen.

The Three Disciplines

Cognitive resilience rests on three disciplines: rotation, redirection, and reflection. Together, they act as firewalls for the mind.

Rotation is the deliberate act of changing perspective to prevent fixation, the silent killer of analytical accuracy. Intelligence services have institutionalized this for decades. During the Cuban Missile Crisis, the CIA formed "Team B" to challenge "Team A's" assumptions about Soviet intent. The aim was not conflict—it was clarity.

The corporate version is the red team. Their purpose is not embarrassment. It is an expansion of thought.

Redirection is the management of attention. When every input competes for urgency, redirection becomes cognitive triage. NASA's mission controllers practice this constantly. When alarms cascade across the console, they use attention discipline to focus only on mission-critical thresholds. Redirection is not a reaction. It is selective depth.

Reflection is a structured pause. It turns exposure into understanding. Naval Special Warfare uses After-Action Reviews after every mission—no exceptions. These are not administrative rituals. They are cognitive reset points.

Reflection forces individuals to reconstruct not only what happened, but why. Reflection is not a delay. It is digestion.

Teams that cultivate rotation, redirection, and reflection develop mental immunity. They stay composed when the tempo of events demands chaos.

The Paradox of Speed

Fast systems demand slow minds. Technology tempts us to equate rapid reaction with expertise, but velocity without reflection breeds fragility. Pilots call this "pilot-induced oscillation"—when overcorrection at high speed amplifies instability instead of restoring control. The same happens in cognition. Overreaction creates oscillation, not insight.

In modern threat environments, autonomous systems analyze millions of signals, and trading algorithms make thousands of decisions per second. Yet the decisions that truly matter— when to escalate, when to intervene, when to trust the system —still unfold at human tempo.

In 2010, the Flash Crash erased nearly a trillion dollars in minutes. Automated trading systems triggered cascading loops faster than humans could respond. Markets recovered, but the lesson was unmistakable. The solution was not faster machines. It was slower humans—circuit breakers, review windows, and enforced pauses.

· · ·

Intentional delay is a cognitive air gap. In cybersecurity operations, forced pauses during escalation and debrief cycles after simulated breaches build resilience by slowing the tempo long enough to extract meaning.

The paradox is evident. The faster our systems become, the more valuable slowness becomes.

Fuel, Firmware, and Friday Night

At 8:17 p.m. on a Friday, the operations bridge for a regional logistics company lit up with endpoint detections from a small distribution center. The agent flagged suspicious encryption. File names were mutating in a temp directory. Network tools showed spikes between a warehouse workstation and the finance server. The site manager called the engineer. The engineer alerted the incident lead.

The pattern suggested ransomware prep. It also resembled a known bug in a recently installed label-printer driver. Comms pushed for a reassuring statement. Legal argued for silence. Operations leadership wanted a go/no-go on isolating the site —an action that would slow overnight sorting.

The incident leader created a three-hour window with a ninety-minute midpoint update. The first thirty minutes went to containment. Segmentation rules blocked traffic to the finance server but allowed printing. Identity forced reauthentication. Backup logs were checked. Snapshots looked clean.

Samples were sandboxed. At ninety minutes, the team reported no ransom artifacts, no persistence, no lateral movement. The printer driver appeared to be generating the file churn. Two details remained unresolved: a suspicious scheduled task and a newly issued vendor hotfix.

The team held posture. After two hours, they pulled the workstation from the network, reimaged it, and rolled back the printer firmware. The scheduled task disappeared. Sorting continued with only minor delay.

Comms released a bounded update at the ninety-minute mark: systems were under precautionary containment, deliveries were operating, and there was no evidence of data exposure. Legal signed off because the message stayed factual.

The incident ended with an answer, not a narrative. A misconfigured driver generated noise resembling ransomware, and a dormant scheduled task triggered maintenance routines. The team added a new control: firmware updates only during midweek mornings with monitored tails.

· · ·

The lesson was direct. Use decision windows to balance speed with certainty. Publish what you know, what you are doing, and when you will say more.

Security of the Mind

Security has always been a process. Now the process begins inside the mind. We spent decades hardening networks and fortifying facilities. The next frontier is cognitive security.

The adversary is no longer just a hacker or a spy. It is fatigue, bias, distraction, and overconfidence. The weapon is not malware. It is momentum—the relentless pull of reaction over reflection.

Cognitive countermeasures require culture, not policy. Organizations that reward reflection over immediacy, dissent over compliance, and understanding over output build stronger mental firewalls than any software.

Elite intelligence units and high-performing corporate teams now formalize cognitive security practices: deliberate pauses to surface assumptions, rotating leadership roles to prevent drift, and embedding behavioral scientists alongside technical analysts. The Israeli Defense Forces' "Team of Ten" model uses multidisciplinary micro-groups that challenge each other's reasoning under time pressure. Their strength comes not from hierarchy, but from cognitive humility.

· · ·

The secure mind is not the one that never errs. It is the one that never stops examining errors.

Security is still a process, not a product. Today, that process begins with attention, discipline, and clarity of thought.

The Cycle of Clarity

Clarity decays. Every moment of understanding erodes as data shifts and attention drifts. The disciplined analyst accepts this. Clarity must be rebuilt continuously.

The cycle is simple: observe → interpret → decide → reflect → recalibrate → repeat.

This mirrors the OODA Loop, first formalized by Air Force Colonel John Boyd. Boyd showed that advantage came not from reacting faster, but from refreshing orientation faster. The one who updates understanding more effectively wins the engagement.

After the 2011 Fukushima disaster, Japan's emergency response shifted from rigid command to iterative clarity. Leaders held rolling situational updates every two hours. It wasn't speed that saved lives. It was a disciplined recalibration.

· · ·

Clarity, like security, is a process—not a product.

Chapter 8

The OODA Loop and
the Art of Clarity

The Mind of John Boyd

Colonel John Boyd never fitted neatly inside the Air Force
hierarchy. He was brilliant, impatient, and allergic to dogma.
To many superiors, he was a disruptor who refused to accept
that tradition equaled truth. To his peers—and later to
generations of strategists—he became something more
enduring: the thinker who learned to find order inside chaos,
the signal inside the noise.

Boyd's journey began in the cockpit. As a young fighter pilot
and instructor, he made a wager that he could defeat any
opponent in simulated combat, even if they began with an
advantage. His record was flawless. That confidence came not
from bravado but from the way his mind worked. He noticed
patterns others glossed over. He read intent in the tilt of a
wing or the angle of a climb. He anticipated the moment
before it arrived.

. . .

He turned that instinct into a theory. His Energy-Maneuverability framework reshaped aircraft design and laid the foundation for the F-15 and F-16. Yet his most significant contribution came not from aerodynamics but from cognition. He realized that survival in any domain—air combat, leadership, or strategy—depends on one thing: the ability to observe, orient, decide, and act faster and more accurately than one's adversary. The OODA Loop was born.

The Orientation Engine

At first glance, the OODA Loop appears simple: Observe, Orient, Decide, Act. Four tidy verbs fit for a slide deck. Boyd's original model was neither neat nor linear. He viewed the loop as a living system—a cognitive engine, not a circle.

Observation delivered raw input. Action delivered visible output. Orientation was the core. It was the part that mattered, the part most people misunderstand. Orientation blends experience, training, cultural background, analysis, and intuition into a constantly updated frame of meaning. When people fixate on "acting faster," they miss the point. The real advantage lies in refreshing orientation faster than the environment—or the opponent—can shift.

Boyd's famous line, "operate inside the adversary's decision cycle," was never about frantic motion. It was about induced confusion. If you can adapt faster than your opponent can comprehend, you force them to rebuild their understanding continually. Their sense of coherence fractures while yours stays intact. Speed matters only because it amplifies the pressure on the slower thinker. The actual weapon is clarity.

The Snowmobile Problem

To teach this, Boyd used his "snowmobile" thought experiment. Imagine a tank, a boat, a bicycle, and a pair of skis. Each works in its own domain. None can solve a mountain-snow problem. But break them apart—take the treads of the tank, the engine of the boat, the handlebars of the bicycle, the skis themselves—and suddenly you can combine them into something new: the snowmobile.

The exercise was a metaphor for creative synthesis. Innovation rarely comes from invention out of nothing. It comes from rearranging the pieces of reality into new configurations. That, too, is orientation: the ability to recombine fragments faster and more insightfully than others.

People who can reorient quickly do not just respond; they act. They redefine the game.

From Dogfights to Doctrine

The Marine Corps adopted it as the foundation for maneuver warfare: tempo, adaptability, and decentralized decision-making over rigid hierarchy. But its reach now extends far beyond combat.

Corporate security teams face a flood of sensor data. Intelligence units sort signals from noise across thousands of data points. Crisis managers operate in an environment of uncertainty as conditions shift by the minute. In every case, the differentiator is not who has the most data. It is those who can refresh their orientation fast enough to recognize weak signals before they become hard shocks.

Boyd's Energy–Maneuverability metaphor still applies. A fighter's "energy" is its ability to maneuver without losing control. For an organization, that energy is cognitive and cultural: the diversity of perspective, the willingness to test small reversible moves, and the humility to discard a view when the facts change.

Healthy organizations execute what Boyd called "fast transients"—small shifts that reveal new information. In a security context, that may mean testing a new escalation workflow, running a red-team drill, or adjusting alert thresholds to see which patterns surface. Every experiment becomes an observation that feeds the next orientation.

The Leader's Loop

For security leaders, Boyd's philosophy offers a practical rhythm:

1. **Observe:** Gather inputs across technical, human, and environmental channels.
2. **Orient:** Reinterpret them in light of the evolving context.
3. **Decide:** Choose the following reversible action.
4. **Act:** Execute and capture what the action reveals.

Then repeat—faster and more coherently than the turbulence around you.

The core question is: after each cycle, what did we observe, how did we interpret it, and what did that reveal about our blind spots?

Organizations that make this habitual develop what Boyd called "repertoire flexibility," the ability to handle unfamiliar challenges without paralysis.

The Bureaucratic Headwind

Boyd understood how fragile adaptability can be. Bureaucracies prefer lines to loops. They reward consistency, not curiosity. They punish uncertainty, not complacency. Orientation slows when dissent is discouraged, and leaders prefer clean narratives over evolving ones.

Boyd spent much of his career fighting not adversaries in the air but institutions on the ground. He was trying to teach rigid systems to think in spirals rather than squares. That friction defined his life. It also made his ideas durable. The OODA Loop endures because it re-centers strategy around cognition rather than compliance.

Automation and the Modern Loop

In today's environment, automation has accelerated the Observe and Act phases beyond human tempo. Machines detect anomalies in milliseconds and respond at machine speed. But machines do not orient. They cannot synthesize ambiguity into insight or weigh ethical tradeoffs under pressure.

Orientation remains human—and it is where leaders must focus their energy. The challenge is not to outrun the machine but to guide it. If observation and action become mechanical while orientation becomes hollow, the loop collapses into automation without judgment.

The OODA Loop becomes not a tactic, but a discipline of awareness. It is a method for continuous sensemaking inside environments that resist clarity.

The Enduring Lesson

Boyd's story is a reminder that innovation begins with defiance. He refused inherited truths. He proved that clarity, not conformity, defines mastery. His loop endures because it captures the essence of adaptation in a noisy world.

To find the signal is to reorient faster than the noise can rearrange itself. To act with precision is to see the pattern beneath chaos. To lead in the modern age is to keep the loop alive—to stay curious, skeptical, and willing to redraw the map.

Chapter 9

Vignette: The OODA Loop in Real Time

The Loop in Everyday Life

The first storm warning came just after sunrise. The coastal sky was calm—gray, quiet, almost gentle. A few fishing boats still pushed toward the horizon. Inside the small emergency management office, the radio crackled with the familiar uncertainty of early forecasts: a low-pressure system had shifted, models were diverging, and the storm now had a name.

By mid-morning, the data feeds conflicted. The National Weather Service predicted one track, a private meteorological firm another. Some models showed a glancing blow. Others showed a direct hit. The mayor wanted a definitive answer. Clara, the town's emergency coordinator, knew there was no such answer.

· · ·

That was the first test of the loop: observation—seeing without assuming. The numbers were inputs, not conclusions. Clara kept watching radar and listening for the tone beneath the updates: the pauses, the hedges, the places where forecasters betrayed their uncertainty.

Orientation came next, though not in the clinical sense most diagrams suggest. Clara stepped away from the screens and looked outside. The wind had begun to turn inland earlier than expected. She called the harbor master. He told her the tide was running against the current, "like it's confused." That detail was enough. Models were probabilities. Pattern recognition came from experience.

She gathered her team and outlined what she knew, what she suspected, and what remained unknown. "We don't have certainty," she said, "but we do have time." Preparations began—not alarms or announcements, but reversible actions: moving vehicles uphill, checking generators, calling nursing homes. Each small step was meant to teach them something.

By late afternoon, the pressure dropped again. The storm had veered west, confirming her instinct. Clara ordered an early evacuation of the waterfront. A few officials resisted; they didn't want to look panicked. She reminded them that waiting for perfect information was just another form of paralysis. "We'll act," she said, "and adjust if we're wrong."

. . .

By nightfall, the storm had intensified beyond any model's prediction. The surge swallowed the first row of docks but stopped short of the hospital ridge. The evacuation saved lives. Power was out across half the town, but no one was trapped or missing.

The next morning, Clara returned to the operations room. Screens were blank, but her notebook was full—timestamps, observations, questions. She was already studying the cycle: what they saw, when they decided, how their orientation shifted, and what they'd refine next time.

It had been the OODA Loop in its purest form. Observe the real, not the expected. Orient faster than confusion spreads. Decide despite uncertainty. Act in small steps that reveal the truth.

This story isn't about the weather. It's about how people think when noise increases. Boyd would've recognized it instantly: a mind finding signal before anyone else could hear it.

Reflection: Composure Inside Chaos

What Clara demonstrated wasn't instinct alone—it was disciplined perception. Her actions reflected a form of intelligence Boyd understood long before dashboards replaced judgment. She wasn't faster than the storm. She was quicker to understand what the storm meant. That distinction is the essence of the loop.

. . .

Observation is not watching more screens. It is noticing what matters before deciding why it matters. Clara didn't cling to forecasts; she watched behavior. She heard unease in the harbor master's voice and trusted that properly grounded intuition is its own data stream. Done well, observation slows time. It creates a gap between signal and story—the gap where clarity enters.

Orientation fills that space. It is the merging of new information with lived experience. Clara's orientation came from years of storms, false alarms, and patterns learned the hard way. Orientation grows each time the loop turns; it is how we stay aligned with reality while it shifts.

From that foundation, decision and action become extensions of awareness. Clara didn't gamble on certainty; she ran small experiments. Each action preserved options and revealed more about the truth. Her strength wasn't command—it was calibration.

When the storm hit, the loop closed. Their choices worked not because they were perfect, but because they were coherent. The team maintained a shared sense of reality as the environment around them changed. That coherence separates poise from panic.

. . .

The story captures the human center of Boyd's philosophy. The OODA Loop is not about aggression or speed. It is about grace under pressure. It teaches that awareness must outrun fear and that clarity is earned through deliberate attention— not through more data.

Whether the crisis is a hurricane, a cyberattack, or a high-stakes ethical call, the same principle holds: the leaders who succeed are those who can stay mentally agile amid turbulence, observe reality without distortion, and act without being consumed by noise.

Boyd's insight endures because it touches something elemental, thinking clearly when the world demands speed. Clarity is not a luxury. It is a survival skill. And in the storms of modern life—data storms, media storms, political storms—it is the one signal that still matters.

From Loops to Lattices

Boyd's OODA Loop was born in the age of jet engines and dogfights. Yet the cognitive principle extends far beyond the cockpit. It explains how individuals, teams, and entire organizations perceive, interpret, and adapt under pressure.

But something fundamental has changed since Boyd's time. The human loop is no longer alone.

· · ·

Machines now observe, classify, and act alongside us—processing inputs at scales humans cannot match. The architecture of cognition has shifted from a single loop to a lattice: interlocking systems of human and machine intelligence, each shaping the other.

This entanglement carries both promise and risk.

Automation expands the field of awareness. It can surface weak signals, detect patterns too subtle for the human eye, and provide early warning long before risk becomes loss.

But the same systems can distort orientation. They amplify bias, flatten context, and tempt leaders to mistake computation for comprehension. The loop can become lopsided—fast at the edges, hollow at the core.

The next frontier of intelligence will be defined by how well we manage this partnership. The question is no longer whether machines can think. It is whether humans can remain oriented while thinking with them.

The challenge ahead is not to outrun artificial intelligence. It is to integrate it without surrendering meaning.

．　．　．

In the chapters that follow, we will explore how adaptive intelligence must evolve to preserve clarity inside this expanding lattice. We'll examine how signal detection, moral reasoning, and the architecture of trust must shift in an era in which observation is cheap, orientation is fragile, and the loop itself has become collective.

The lesson of Boyd still holds. Those who can reorient fastest will shape the future. The difference now is that orientation is shared—with the machines we built and the uncertainty we have yet to master.

Chapter 10
The Architecture of Signal

Every organization has a signal architecture, whether it recognizes it or not. It rarely appears on an org chart or workflow diagram, but it exists nonetheless. It is the unseen framework that determines what gets noticed, what gets ignored, and what becomes "real" inside the system. It is the nervous system of collective awareness—the channels and filters through which perception travels, and meaning takes shape.

The Red Folder Test (Ford v Ferrari, 2019)

Shelby sits across from Henry Ford II. He mentions the red folder on the CEO's desk and notes that he watched it pass through four pairs of hands before reaching Ford. In that one moment, the film shows how incentives, gatekeepers, and layers of interpretation shape what a leader sees. Shelby's request is not just about resources; it's about rewiring how truth moves. He wants a direct lane to reality, unfiltered by politics.

. . .

What it reveals: Important facts were delayed and diluted at every handoff. The org chart created a distance between the ground truth and the decision-maker. The change that mattered wasn't messaging—it was *architecture.*

Trace one critical decision from source to sign-off. Count the handoffs. Note what changed at each step. Identify where raw signals should bypass intermediaries and reach the accountable owner in their original form. Build a direct lane for frontline truth with clear criteria, timelines, and ownership.

Every organization already has a signal architecture. If you don't design it, it will make your decisions for you.

The Architecture That Builds Itself

Most organizations do not deliberately build their architectures. They evolve through drift. Over time, tools replace judgment, metrics replace meaning, and reports become stand-ins for reflection. What begins as an information flow calcifies into a *hierarchy of interpretation.*

Organizations start operating not on reality, but on what their systems are designed to see.

. . .

The Deepwater Horizon disaster stands as a harsh example. Pressure anomalies and equipment irregularities were known on the rig, yet the reporting structure prioritized compliance over weak signals. The indicators didn't vanish—they were buried under architecture.

Efficiency replaced awareness. The system was optimized for the wrong thing.

An intentional architecture reverses that entropy. It does more than move data. It moves *meaning*—without distortion, without delay, or drowning the recipient in noise.

The Architecture of Perception

Signal architecture is not primarily about technology. It is about design philosophy. The governing question is simple:

What does this organization want to see clearly?

Everything follows from that answer—dashboards, workflows, meeting cadence, reward structures, blind spots. If an organization measures volume, it manufactures noise. If it measures relevance, it cultivates discernment.

After the 1977 Tenerife runway collision, aviation authorities didn't just retrain pilots—they redesigned cockpit communication. Hierarchy became structured dialogue. Any crew member could challenge assumptions. They didn't revise a skill. They revised *perception*.

Every tool reflects a worldview. Dashboards don't just display reality; they construct it. They dictate what counts as signal and what dies as noise.

Leaders who fail to recognize this invite entropy. Leaders who understand it become architects of awareness.

Curation as Discipline

Curation is not denial. It is discipline.

It defines the boundaries of attention, protecting cognitive bandwidth so teams can see what matters. In a culture obsessed with "total visibility," curation feels countercultural—but total visibility is a myth. It produces only shallow comprehension and chronic overwhelm.

Good curation prioritizes interpretation, not accumulation.

After 9/11, the intelligence community realized the problem wasn't a lack of data. It was a lack of structural coherence. The creation of the National Counterterrorism Center wasn't a data grab—it was an architectural correction.

· · ·

Curation turns fragments into a pattern and patterns into perspective. In an era when everyone sees everything, the advantage belongs to those who know what to ignore.

Adaptive Intelligence Cells

Intelligence doesn't scale by bulk. It scales by structure.

Highly effective architectures depend on small, independent teams that are aware of their context—these are called "intelligence cells." They operate like neural clusters, each responsible for sensing and interpreting a specific slice of reality.

JSOC's transformation in Iraq captured this perfectly. McChrystal dissolved hierarchical silos and built a distributed network in which teams constantly shared insights. Tempo increased tenfold without losing clarity.

The biological metaphor holds: distributed perception yields resilience.

But resilience only emerges when two forces coexist:

- **Autonomy of judgment** (freedom to interpret)
- **Alignment of purpose** (shared mission, shared priorities)

Autonomy without alignment becomes chaos. Alignment without autonomy becomes blindness. Leadership is the dynamic balance between the two.

These cells aren't anti-hierarchical. They're anti-stagnation. They keep the system alive.

Leadership as Architecture

Leadership at its highest form is architectural. Every policy, channel, ritual, and review is a design element that shapes how perception flows.

Most leaders manage information. Exceptional leaders design interpretation.

Mossad institutionalized "critical friend" reviews. Teams challenge each other's assumptions before making decisions. Not as critique—but as calibration.

Amazon replaced decks with written narratives, not for aesthetic reasons, but because writing fosters deeper reasoning and shared comprehension.

Leadership as architecture means building systems that slow thinking just enough to be accurate—and accelerate insight just sufficient to be decisive.

· · ·

The best leaders are not omniscient. They are architects of inquiry.

The Intentional System

An intentional signal architecture transforms organizations from reactive to reflective. It builds a rhythm of comprehension—a cadence through which complexity is metabolized into clarity.

Its principles are simple but profound:

- Curate for comprehension.
- Distribute intelligence through adaptive cells.
- Align through shared trust and language.
- Lead as a designer of perception, not a controller of data.

When these converge, the organization becomes self-correcting. It spots distortion early. It interprets anomalies with context. It learns faster than conditions can change.

This is the new definition of resilience: not just defense, but disciplined awareness.

Every organization gets the awareness it designs for. Signal architecture is the blueprint of that awareness.

· · ·

The question is never whether an architecture exists, but whether it was built intentionally or allowed to happen.

Signal architecture defines perception, but perception alone cannot hold a system together. Systems built for awareness still require belief to stay coherent. Trust is the geometry that binds perception into shared meaning.

The next chapter explores how belief travels through systems —how trust anchors interpretation and can be reinforced by design. No signal endures without trust in the network that carries it.

Chapter 11
Clarity Architects:
The Moral Perimeter

Real World Examples

Every era produces a small number of people who refuse to accept the status quo. They look at the turbulence of their time and sense a deeper pattern—something faint but coherent beneath the surface. They do not worship equilibrium; they go toward the places where equilibrium breaks. These are the builders of clarity. They find the signal long before others even recognize the noise.

Their work, spread across science, industry, and exploration, shares one trait: they learned to interpret instability rather than fear it. In their hands, disorder was not a threat but a diagnostic. They treated uncertainty the way John Boyd treated air pressure or a fighter pilot's angle of attack—as information.

. . .

This chapter is about those interpreters—the ones who shaped entire disciplines by developing a new relationship to ambiguity, speed, and change.

Ilya Prigogine: Order Born from Turbulence

Where Boyd studied the agility of thought under pressure, Ilya Prigogine studied the agility of nature. Both men approached chaos not as failure but as raw material.

In most of modern science, disorder was part of an equation that scientists sought to minimize. Prigogine did the opposite. He walked toward it. He asked a question that unsettled his peers: *What if instability is the source of organization?*

His answer became the theory of dissipative structures— systems that do not collapse under stress but instead reconfigure themselves into higher forms of order. In Prigogine's world, turbulence was not decay; it was the engine of evolution.

Translated into the language of intelligence and security, Prigogine's insight is blunt: The systems that survive are the ones that metabolize volatility. Everything else gets left behind as the environment changes.

He treated chaos as a teacher. Boyd would recognize that immediately. Both men understood that orientation—the

ability to absorb change and turn it into understanding—
determines survival.

The moment in *Interstellar* when Cooper enters the black hole
captures this visually: everything breaks, light bends, time
unravels. Yet inside that collapse, a structure appears. Not
stability, but intelligibility. Prigogine spent his career showing
that this is how the universe works: the closer you move to the
edge of disorder, the more you learn what the system can
become.

Taiichi Ohno: Teaching Systems to See

If Prigogine found a pattern in turbulence, Taiichi Ohno found
it in constraint.

Walking the factory floors of Toyota in the 1950s, Ohno saw an
environment drowning in noise—inventory, movement, waiting,
delays. To others, these were the standard conditions of
industrial life. To Ohno, they were symptoms of a system with
poor eyesight.

He introduced a new discipline: Don't optimize the machine.
Improve the perception.

From that came the Toyota Production System: a philosophy
built on immediacy of signal, tight feedback loops, and clarity
through flow. Ohno treated every production line as a living

sensor. A defect wasn't a failure; it was a message. A delay wasn't an inconvenience; it was a weak signal about structure.

This mindset is pure Boyd:

- Observe the environment without filters.
- Orient by exposing assumptions.
- Decide quickly, locally, and reversibly.
- Act in small steps that improve understanding.

Ohno's genius was not efficiency. It was perception engineering—teaching organizations to see problems early enough that they cost almost nothing to solve.

The cinematic parallel is *The Martian*. Watney survives not by brilliance but by iteration: try, observe, adjust, try again. That cadence—fast transients, continuous learning—is the core of Ohno's system and the reason it reshaped industries far beyond manufacturing.

He proved that clarity is not a product of more data. It is the product of better attention.

Fridtjof Nansen: Adaptation at the Edge of the Map

Where Prigogine studied nature, and Ohno studied systems, Fridtjof Nansen studied people at the limits of both. He spent his life moving between frontier environments—glaciers, failed states, famine-stricken regions—and he learned the same truth in every one of them:

When the environment no longer aligns with the map, leadership becomes improvisation.

During his Arctic expeditions, Nansen discovered that survival depended on interpretation, not force. When his ship was trapped in ice, he did not fight the drift—he studied it. The ice became a kind of signal, carrying his expedition across the polar basin. He navigated by reinterpreting the constraint. His OODA Loop was written in wind, water, and silence.

Most explorers stop there. Nansen did not. After World War I, he stepped into a different kind of frozen landscape—millions of displaced people with no legal identity. Bureaucracy had become ice: rigid, slow, impersonal.

Nansen did what Boyd or Ohno would have done: He redesigned the system. He created the Nansen Passport, recognized by multiple nations, that enabled stateless people to move, work, and survive.

· · ·

He treated humanitarian collapse the way he treated Arctic storms—not as tragedy alone, but as a problem of orientation. People needed a new mechanism to navigate a world that no longer matched their expectations. He built it.

The closest cinematic parallel is *Apollo 13*: the engineers tasked with making one set of tools fit another, with no time and no margin. Nansen lived his entire career inside that constraint. He engineered pathways where none existed.

Nansen's contribution to this book's core argument is simple: Clarity without empathy is brittle. Orientation requires not just perception but purpose.

The Thread That Connects Them

Prigogine, Ohno, and Nansen worked in different worlds— thermodynamics, industrial design, polar exploration, and humanitarian law. But their disciplines converge on one principle:

The systems that endure are the ones that turn instability into insight.

They treated uncertainty as a condition to be understood, not as something to be denied.

They built architectures that:

- Reveal weak signals early.
- Reward attention over assumption.
- Reorient faster than the environment
- Transform turbulence into comprehension.
- Place human judgment at the center of adaptation.

What Boyd accomplished through maneuver theory, these figures accomplished through science, industry, and diplomacy. Each expanded the OODA Loop to their domain.

And each demonstrated something the modern security environment desperately needs to relearn: Innovation is not speed. It is orientation. Not acting faster, but understanding deeper and not predicting the future, but adapting to what unfolds.

They are reminders that in every era of volatility, a few people find the same faint signal:

- Chaos is information.
- Friction is feedback.
- Uncertainty is energy.
- And clarity is a discipline, not a gift.

The next chapter turns from these individual exemplars to the organizational implications: how leaders can build environments that produce this type of clarity not once, but continuously.

The world is not getting quieter. But clarity is still possible when the architecture is exemplary.

Every expansion of sight expands power. From the watchtower to the drone, from the magnifying glass to the algorithm, vision has always carried consequences. To see more is to alter what is seen. Yet in the modern signal age—where sensors are ambient, analytics are autonomous, and data flows without friction—the connection between perception and responsibility has frayed. We observe automatically, interpret through proxies, and act on inferences without acknowledging that **observation itself is an intervention.**

The discipline of intelligence has always rested on a tension: the pursuit of understanding versus the temptation of exploitation. The boundary is defined not by capability, but by restraint. Without restraint, visibility mutates into surveillance. Without ethics, insight dissolves into manipulation.

The central question is no longer *what we can know. What should we do with what we know?*

This chapter confronts that moral perimeter.

Restraint as Architecture

The most dangerous illusion of modern intelligence is thinking that ethics is an add-on—a policy appendix, an endnote, a compliance box. Ethics is architecture. It shapes how information moves, how it is interpreted, and how much power the observer is allowed to wield.

The Cambridge Analytica scandal in 2018 made this painfully clear. The violation was not data collection; companies had been harvesting behavioral traces for years. The violation was intentional. Information meant for personalization was retooled for persuasion. The perimeter collapsed.

Information without limits becomes intrusion. Insight without integrity becomes exploitation.

Ethical intelligence is not about avoiding uncertainty; it is about navigating it with discipline.

The Apple–FBI encryption dispute in 2016 showed a different posture. Apple's refusal to create a universal backdoor was not technical defiance—it was architectural restraint. The company understood that extending its line of sight for one case would permanently alter the social contract for millions. Restraint was not weakness, but stewardship.

. . .

This is the nature of the moral perimeter: a boundary built not from ignorance, but from principled limitation.

The Fox Clause (The Dark Knight, 2008)

Batman builds a phone-sonar system capable of seeing every corner of Gotham. He gives control to Lucius Fox, who immediately refuses: "This is too much power for one person."

Batman answers, "That's why I gave it to you," and hard-wires the system to self-destruct once the mission ends.

It is not the technology that makes the system ethical. It is the sunset clause, the ownership, and the restraint.

Extraordinary sensing requires extraordinary stewardship. Governance is a design requirement, not an afterthought.

For any high-sensitivity system, identify a human custodian, define use rules, set time limits, and integrate a kill switch into the system's architecture. Auditability and access limits matter as much as accuracy.

The Age of Cognitive Colonization

The frontier of intrusion is no longer physical. It is cognitive.

· · ·

When algorithms shape perception before we consciously choose, the environment becomes a form of preemptive influence. Not coercion but curation. Not manipulation but momentum.

This is cognitive colonization. When systems decide what we think *about* long before we decide what we believe.

The Facebook Papers showed how easily this happens. Algorithmic amplification rewarded emotional charge over nuance because outrage increases watch time. The result was synthetic polarization—engineered not through deception but through design.

People are not being watched. They are being shaped.

Ethics in this environment is not optional. It is existential.

Without moral counterweights, intelligence systems become engines of perceptual capture. They shrink the range of interpretation. They dull skepticism. They nudge cognition toward predictable, profitable grooves.

The threat is not that machines see us. It is that they begin to think on our behalf.

The Discipline of Moral Design

The antidote to cognitive colonization is moral design—a shift from ethical rhetoric to ethical engineering. It treats principles as infrastructure.

Moral design begins with uncomfortable questions:

- What are we optimizing—understanding or influence?
- Whose interests do our metrics serve?
- At what point does pattern detection become behavioral control?

These are operational questions, not philosophical ones.

The General Data Protection Regulation (GDPR) scaled this idea. Its requirements—transparency, necessity, consent—forced organizations to justify their data practices. Not all companies liked it, but the regulation reframed ethics as a structural constraint rather than an aspirational value. It made restraint measurable.

Actual moral design accepts this asymmetry: the observer always has more power than the observed. Restraint is how we equalize that imbalance without dulling insight.

Mirrors and Smoke

At 7:43 a.m., the controller at a mid-size firm receives a call from a voice indistinguishable from the CEO's, with the same tone, cadence, and urgency. A confidential acquisition is closing. An eight-million-dollar wire must go out immediately.

The email that follows looks plausible. The deal name is familiar. The forwarded chain includes counsel. The headers pass a casual glance; pressure to act mounts. Familiarity becomes a vulnerability.

The finance team uses a shared-secret protocol for high-value wire transfers. Verification requires a callback to a stored number and a challenge known only to leadership. The controller holds the line. The callback exposes the deception. A deeper header review reveals a domain spoofed with Unicode lookalikes.

No funds move. A companywide note reinforces the protocol: callback, shared secret, and header analysis. Finance adds a new requirement: public approval from a second executive in Slack for all urgent wires.

When perception can be manufactured, do not trust familiarity—trust friction. Ethics is not only about what we see, but also about how we decide what to believe.

Transparency with Boundaries

The instinctive response to eroded trust is to be radically transparent. But transparency without boundaries becomes exposure. The equilibrium lies between the two.

After the Snowden disclosures, the U.S. intelligence community began publishing transparency reports— summaries of surveillance activity without compromising sources. This partial openness rebuilt credibility without weakening security. People did not need every detail. They needed to understand the decision logic behind the collection and its rationale.

Transparency is not confession; it is coherence. It is how organizations communicate the architecture of trust.

The Ethics of Restraint

The highest form of intelligence is restraint.

The ability to hold knowledge without abusing it. To see vulnerability without exploiting it. To understand without steering.

Restraint is the immune system of intelligence. It prevents clarity from mutating into coercion. It protects both the watcher and the watched.

. . .

Organizations that internalize restraint—through governance, culture, and deliberate limits—build legitimacy into their signal systems. The best modern AI labs practice this explicitly: red-teaming models, enforcing ethical review gates, and designing stopgaps into deployment pipelines.

Ethics is not a brake on innovation. It is the condition that allows innovation to be trusted.

Power with Humility

Seeing clearly is dangerous without humility. Every dataset is a record of a human life. Every pattern is a story only partially told. Every act of perception carries a moral shadow.

Edward R. Murrow warned in 1958 that technology could "teach, illuminate, and inspire—*but only to the extent that it is used with conscience.*" That warning defines the modern signal age. The problem is not visibility. It is the failure to design for responsibility.

Humility is the final perimeter. It is what keeps intelligence from becoming domination. It is what keeps clarity from becoming coercion.

. . .

We now see more than any previous generation. Whether that vision leads to understanding or to intrusion will depend on one discipline: the courage to observe with restraint.

The signal age has given us extraordinary sight. Only ethics can determine what we choose to see.

Chapter 12
Simulated Certainty

Noise is multiplying. Each day, the line between authentic and artificial erodes a little more. Synthetic data, deepfakes, and algorithmic hallucinations flood our environment with convincing forgeries. What once resembled a shared map now feels like a contested battlefield. Coherence no longer guarantees accuracy. Confidence can be counterfeit.

Verification used to be a supporting act. Now it is the main struggle. We are entering an age of simulated certainty, where falsehoods arrive pre-polished and almost indistinguishable from fact. The tools that generate this noise are not inherently malicious. They are indifferent. Their danger lies in scale.

In 2023, a viral image of Pope Francis in a white Balenciaga puffer coat raced across social media. Millions accepted it as real. The image was generated in seconds by an AI model. No hostile state actor, no coordinated campaign. Just a convincing illusion, accelerated by curiosity and humor.

That incident showed something important. Indifference at scale can be more destabilizing than malice in isolation.

The next frontier of intelligence will not be speed or reach. It will be discernment under deception—the ability to maintain composure and judgment when the world itself can be generated adversarially.

Synthetic Realities

Deepfakes, generative models, and synthetic identities not only threaten facts but also threaten the very fabric of reality. They threaten trust.

When reality can be fabricated with the precision of memory, skepticism becomes both necessary and exhausting. Jean Baudrillard warned that in a "hyperreal" world, simulations do not simply imitate reality. They begin to replace it. We are approaching that edge, an epistemic event horizon where information struggles to prove itself.

When the Fake Feels Truer Than the Real

In *The Truman Show*, the architect of Truman's world says, "We accept the reality of the world with which we're presented." The set works because it is seamless. It does not just distort facts. It creates a seamless environment — and the brain relies on coherence.

In *Spider-Man: Far From Home*, Mysterio weaponizes illusions and drones to stage crises for the cameras. "People need to believe," he says. "And nowadays, they'll believe anything." His power is not only technical. It is psychological. He understands that if you present the truth persuasively enough, institutions will treat it as real.

Deepfakes and synthetic media are dangerous because they fabricate entire contexts, not just isolated images. They make the fake seem more believable than the real. That is the actual risk of hyperreality.

In this environment, authenticity becomes both the most valuable and the most vulnerable asset. The institutions that endure will not be the ones that hoard data. They will be the ones who preserve context. That means:

- For journalists: traceable sourcing and public-verification workflows.
- For corporations, securing not only information but the trust frameworks around it.
- For intelligence agencies, authenticating not just signals, but their provenance.

When a deepfake video of Ukrainian President Volodymyr Zelensky appeared in 2022, urging soldiers to surrender, the response was instructive. Officials countered quickly, not by trying to erase the video, but by publishing authentic footage and explicit confirmation that the clip was false.

In an age of synthetic realities, credibility becomes a form of defense. Verification is no longer bureaucratic. It is existential.

The defining skill of the next decade will not be collecting data. It will be authenticating it. Verification will become an act of leadership.

Cognitive Sovereignty

The new strategic advantage is cognitive sovereignty—the ability to think freely inside engineered influence.

In an era of synthetic certainty, sovereignty begins with thought, not territory. It is the discipline of perception that resists capture, the calm that persists under distortion, and the clarity that survives manipulation.

Cognitive sovereignty cannot be automated. It must be trained, protected, and renewed. It is a team's posture: recognizing distortion without absorbing it. The leader who can hold ambiguity without panic, and the analyst who can filter bias without burning out, embody this discipline.

. . .

The history of deception makes this clear. Operation Fortitude, the Allied campaign that misled Germany about the D-Day landing site, succeeded because Allied planners controlled their own cognition while deliberately flooding the enemy's. They could think clearly inside their own fog.

Today, disinformation campaigns target entire populations with that same logic. They do not need to persuade everyone. They only need to exhaust attention until people no longer know what to trust.

The defense is not censorship. It is composure. Cognitive sovereignty is not isolation. It is integrity.

Leadership in the Age of Simulation

In the age of simulation, leadership will mean composure in confusion.

When systems fail and truth fractures into competing versions, the leader's first job is not to act. It is too steady. The trait that will carry the highest premium is not raw decisiveness. It is equanimity under pressure.

When the U.S. airline system's Notice To Airman (NOTAM) network failed in January 2023, grounding flights nationwide, speculation spread faster than information. The leaders who restored trust were the ones who resisted the urge to improvise explanations. They communicated calmly, acknowledged uncertainty, and shared verified details only as they became available.

Their tone became the signal.

In high-stakes environments, from crisis communication to cyber incident response, the leader's nervous system becomes the final trusted channel. A steady voice can restore coherence when information itself is unstable.

In that sense, psychological steadiness is a strategic infrastructure.

Hearing Through the Static

The world will not grow quieter, but we can relearn how to hear.

In an era of algorithmic saturation, deep listening becomes a competitive advantage. It requires humility. The signal may not arrive as a clean, labeled input. It may show up as dissent, contradiction, or silence.

. . .

Organizations that listen deeply across hierarchy and geography will detect what automation misses. The Boeing 737 MAX crisis showed what happens when systems go deaf. Engineers and pilots raised early concerns about control software. Those concerns were diluted by corporate noise. The result was a catastrophe.

By contrast, Toyota and NASA have institutionalized "listening loops," creating environments where junior staff and remote teams are expected to surface anomalies. These organizations do not rely only on dashboards. They treat human voices as sensors.

Authentic listening converts noise into signal. Even in distortion, there is rhythm. Even in chaos, there is cadence. The future art of clarity will not be about eliminating noise but about recognizing the patterns within it.

The Siren and the Silence

At 9:12 a.m., a county alert is sent to mobile phones and highway signs. Chemical release. Shelter in place. The operations team at a nearby campus sees it at the same time as everyone else. The campus includes labs and light manufacturing. Radios fill with questions. Employees text supervisors. A local news station amplifies the alert without context.

· · ·

The alert might be real and relevant. It might be a test misfire. The county's public portal is quiet. The state emergency site shows nothing. On a neighborhood social feed, a photo shows a siren truck. Another shows clear skies over the named plant. An on-site engineer notes that prevailing winds would carry any airborne hazard away from the campus. Risk exists, but it is unclear. The team must decide whether to trigger a shelter-in-place protocol that shuts down air handlers and locks doors.

False activation will erode trust. Failing to identify a real hazard will put people at risk.

The incident creates a 20-minute window and splits the work. One person calls the county emergency coordination center. One calls security at the neighboring plant. One stages the shelter protocol so it can be executed quickly if needed. The communications lead drafts a brief message to employees acknowledging the alert, explaining the response, and providing a specific time for the next update.

At minute eight, the plant's security officer confirms a test had been scheduled for another day. At minute 10, the county call center reports it is investigating a possible cross-feed between test and live systems. At minute fifteen, the county posts on social media that the alert was sent in error. The campus team still waits for a second independent confirmation before closing the window.

. . .

The first internal message goes out at minute five. It outlines what is known, what is unknown, and when the team will provide more details. The second message at minute twenty confirms the false alarm, citing county and plant sources. The campus never triggers a shelter-in-place. Air handlers stay on. Doors remain open. People see a process working openly.

There is no rumor mill, because the team communicates uncertainty without shame. In the After Action Review (AAR), they subscribe to the county's machine-readable alert feed and write a "reverse verification" checklist for public alerts. It lists which sources to check, in what order, and includes templates for the first two messages.

The lesson is simple. Public trust depends less on perfection and more on visible cadence. Silence breeds fear. Overreaction breeds cynicism. Short, honest updates keep confidence intact.

The Enduring Process

Clarity is not a destination. It is a discipline.

It erodes with time and must be rebuilt as information shifts and human attention tires. The future of intelligence and security will belong to those who can sustain the cycle of sensemaking: observe, interpret, decide, reflect, recalibrate, repeat.

. . .

Mature systems already live this way. Aviation's Safety Management System is built on continuous feedback, not one-time assurance. Pilots debrief after every flight, even the uneventful ones, to capture minor deviations before they become major failures.

The lesson is universal. Resilience does not come from certainty. It comes from renewal.

The organizations that endure will not promise stability. They will promise adaptation. Security remains a process, not a product. Clarity, its intellectual twin, must be treated the same way. It has to be cultivated, protected, and renewed as a daily practice.

The world will not be quiet for us. Our advantage will come from learning to quiet ourselves within it, to hear through the static, and to keep finding the signal in the noise.

Chapter 13

The Custodian
of the Loop

"Human in the loop" entered our vocabulary as a promise. A safeguard. A reassurance that even as machines accelerated, human judgment would remain the final arbiter. It was meant to be the ethical seatbelt of automation: no matter how fast the system, a person would always retain the ability to intervene.

In practice, that promise has thinned. Automation has outpaced cognition. Decisions that once took minutes now unfold in milliseconds. Even the most attentive operator cannot meaningfully interject at the speed of system-to-system interaction. The human still appears inside the loop diagram—but increasingly as an observer of choices already executed. The loop remains intact on paper as it dissolves.

We built automation to reduce errors. Often, we only moved the error. Machines do not tire, but they do not reason. They do not understand motive, irony, or context. They cannot distinguish between an outlier and a warning.

In many modern systems, the human is technically present but functionally irrelevant—a custodian of liability rather than a participant in understanding. The result is a new form of responsibility without control: moral accountability for systems we can no longer meaningfully steer.

The Tempe Collision: A Loop Too Fast

The 2018 fatality involving Uber's autonomous test vehicle in Tempe, Arizona, crystallized this tension.

The system repeatedly misclassified a pedestrian—first as a bicycle, then as a car, then as an unknown object—oscillating between labels and delaying action. The safety driver, expected to override the system if needed, was looking away. Years of system reliability had dulled vigilance. The automation failed to orient, and the human reacted too late.

The driver was prosecuted. The code was not.

But the deeper failure was architectural. The role assigned to the human—constant, perfect vigilance amid long stretches of flawless automation—is psychologically unsustainable. When the system is too reliable to stimulate attention yet too brittle to handle the unexpected, oversight collapses. This is automation bias in its purest form: trust overwhelming vigilance.

The human was in the loop, yes. But the loop was moving too fast for the human to matter.

When Automation Overrules Attention

Aviation learned this lesson long before autonomous cars. Air France Flight 447, which crashed in 2009 after its pitot tubes froze, showed the consequence of overdependence on automation. When the autopilot disengaged, the pilots—stripped of reliable indicators and unused to flying manually at high altitude—overcorrected into a stall.

The pilots had oversight in theory. But their skills had atrophied due to automation. Their situational awareness evaporated just when it was needed most.

The NTSB now labels this pattern *automation complacency*: systems become so dependable that they train humans to disengage until the rare moment when the machine fails—and then the human is too slow to recover.

Where Machines Surpass Us

The counterargument is undeniable. In countless domains, machines outperform us:

- Autopilot reduces fatigue-related aviation incidents.
- Cancer-detection algorithms spot anomalies doctors overlook.
- Algorithmic trading optimizes liquidity with inhuman precision.
- Assembly-line robots weld with flawless consistency.

In repetitive, high-volume, low-judgment environments, humans are often the weakest link. Automation eliminates variability. It reduces error.

Until the world behaves in ways the system has never seen.

It is at that rupture—where routine ends and ambiguity begins—that the loop reopens, and the human becomes essential again.

When Context Returns: The Flash Crash

The 2010 Flash Crash, which erased nearly a trillion dollars in minutes, is the clearest example of automation's blind spot. Algorithmic trading systems reacted to one another's movements without understanding meaning or consequence. Volatility amplified volatility. Context disappeared.

Human traders reintroduced it.

· · ·

Some firms slowed trading manually. Others switched algorithms off entirely. The firms that fared best were those that had preserved "manual intervention protocols"—explicit authority for a person to halt automation when numbers stopped resembling reality.

Speed without sense creates fragility. Context, not computation, restores coherence.

Cybersecurity and the Limit of Classification

Cybersecurity amplifies the same dilemma. Automated intrusion detection systems can parse millions of events per second. They surface anomalies no human could detect.

But they cannot assign significance.

A spike in traffic might be an exfiltration attempt. Or a software update. Or a misconfigured backup job.

Only humans understand business context. During the Colonial Pipeline ransomware attack, automation escalated alarms, but only human judgment could weigh operational, reputational, and national impact. The shutdown was a human decision—a high-risk, high-cost attempt to preserve the future at the expense of the present.

· · ·

Automation detected the threat. Humans understood its meaning.

Medicine and the Meaning Gap

In diagnostic imaging, machine-learning systems routinely outperform radiologists at detecting anomalies. Yet radiologists describe a persistent "meaning gap."

AI can identify an irregularity. It cannot decide whether it matters.

It cannot integrate patient history, emotional nuance, family risk factors, or the lived context behind the image. Here, the human's role shifts from operator to interpreter: not re-checking machine outputs line by line, but assembling a coherent picture of a life.

Precision without perspective breeds misunderstanding.

The Custodian of Meaning

This is the essence of the loop dilemma. Automation excels at precision but cannot assign weight. It can highlight deviation but cannot interpret motive.

The human role is not to match the machine's speed. It is to supply what the machine cannot generate: meaning, proportionality, context, and intent.

We are not backup systems. We are custodians of significance.

Designing for Human Relevance

The real question is no longer whether humans should remain "in" the loop, but how to stay relevant within loops dominated by machine tempo.

This requires architectural change:

Expose uncertainty, don't hide it.

Interfaces should reveal ambiguity, not mask it behind confident numbers. Proper decision support surfaces what the model *cannot* know.

Graduated autonomy

Systems must allow humans to reclaim control dynamically— not only after catastrophic failure.

Cognitive workload engineering

Vigilance must be trained, not assumed: micro-interruptions, anomaly drills, rotation of monitoring roles, periodic re-engagement cycles.

Explainability as a design element

Decision support should show lineage, not just output. Why was this alert triggered? Which assumptions drove the model?

NASA's rover teams and the U.S. Air Force's Loyal Wingman program embody this model. They blend autonomous capability with human oversight engineered for interruption and re-engagement. The person is not a spectator. They are a strategic editor of machine action.

The Moral Center of Automation

When the loop breaks, accountability still lands on us. The machine may execute the action, but the consequences always rest with the designer, approver, operator, or leader.

Ethical automation requires not only human presence but human comprehension. Without understanding, responsibility becomes theater. Without interpretation, oversight becomes fiction.

The future of "human in the loop" is not about keeping a person near the controls. It is about ensuring the human remains the center of meaning, the moral anchor against which machine decisions are measured.

. . .

Intelligence without judgment is indifference. Automation without humanity is acceleration without direction. The signal survives only as long as someone is still able to interpret it.

In that role—the interpreter of context, the custodian of consequence—the human remains irreplaceable.

Chapter 14

Cognitive Armor

Stress does not destroy cognition. It distorts it. Under pressure, the mental aperture narrows. Attention collapses toward whatever feels most immediate, while abstraction, empathy, and long-range reasoning retreat to the edges. That narrowing once kept us alive. On the savannah, tunnel vision meant survival. In a command center, an operating room, or a boardroom, it quietly sabotages judgment.

In the signal age, the threat is rarely physical. It is informational. Too much, too fast, from too many directions. We face cognitive predators, not lions: overload, uncertainty, distraction, fatigue. The body responds as though hunted, but the chase never ends. The result is decision fatigue and perceptual drift—the silent failures that unravel complex systems.

· · ·

The antidote is cognitive armor: the discipline that protects thought from distortion. It is not a mindset. It is not optimism. It is a trained reflex—the ability to remain deliberate when the environment demands reaction.

High-reliability domains learned this first.

The Architecture of Deliberate Thinking

Aviation, medicine, military operations, and emergency response all share a core assumption: stress is inevitable, and no level of talent is immune to its effects.

That is why these fields rely on tools that seem almost mundane—checklists, verbal confirmation loops, predefined sequences. They are not crutches. They are prosthetics for memory under stress.

A checklist is not a sign of weakness. It is a recognition of physiology.

Aviation: When Human Factors Became the Real Threat

By the late 1970s, investigators noticed a troubling pattern. Many of the worst aviation disasters—Tenerife, Portland, Cincinnati—were not caused by mechanical failures but by cognitive ones: breakdowns in communication, unchecked assumptions, unchallenged decisions.

. . .

Tenerife remains the clearest example. A captain, under time pressure and miscommunication, began takeoff on a fog-shrouded runway where another aircraft sat unseen. The systems worked. The people did not.

The industry responded with one of the most effective cognitive redesigns in modern history: Crew Resource Management (CRM).

CRM reengineered cockpit behavior. It emphasized shared situational awareness, structured communication, and deliberate calm. It taught crew members to challenge hierarchy, to verbalize doubt, and to treat procedure as protection—not paperwork.

The results were transformative. Within a decade, accident rates tied to human factors dropped across commercial fleets. CRM became the operating system of modern aviation, later adopted by NASA and military aviation units.

It was cognitive armor in its purest form.

Medicine: Checklists as Shields Against Chaos

In the early 2000s, Atul Gawande and the World Health Organization found that many surgical complications stemmed not from skill but from communication failures—a missing tool, an unconfirmed dosage, a misidentified patient.

The fix was a nineteen-item verbal checklist.

Mortality dropped by more than a third. Not because the checklist made surgeons smarter, but because it made their teams coherent when stress was highest.

Experienced surgeons initially resisted. "Real professionals don't need checklists," they argued. But the data won. And the lesson was universal: structure protects cognition. Deliberate pauses convert chaos into coordination.

Corporate Life: Decision Fatigue at Scale

Executives and analysts face the same cognitive forces under different names. During cyber incidents, market shocks, or reputational crises, they confront the same narrowing of attention, the same surge of adrenaline, but without the kinetic cues that signal danger.

One global bank discovered this the hard way after a costly breach. Their response teams defaulted to speed over clarity, producing overlapping actions and contradictory reports. The fix was simple but profound: Cognitive Reset Protocols.

• • •

When an incident escalated, the commander called a structured pause: Laptops shut. Eyes up. Thirty seconds of silence. Then: current facts, assumptions, unknowns.

Over a year, this ritual produced a 25 percent improvement in post-incident accuracy and a significant reduction in duplicate work. The pause had become their cockpit checklist.

Military and Emergency Operations: Training Under Distortion

Special operations units and aviation crews have long treated stress inoculation as core training. Instructors deliberately overload recruits—noise, time pressure, conflicting data—so they learn to execute in the face of chaos.

The Navy SEALs' "40 percent rule" captures the philosophy: when you think you're finished, you are operating at only 40% of your capacity. The remaining sixty percent is blocked by cognitive noise.

Cognitive armor is the discipline that safely unlocks it.

Firefighters, paramedics, and urban search-and-rescue teams use similar tools:

- Command time-outs,
- Structured call-and-response,

- After-action reviews within minutes of an event.

Composure is not natural. It is engineered.

The Counterargument: The Romance of Instinct

Critics warn that structure kills creativity. They argue that instinct saves lives and procedure slows insight. There is truth here: over-structuring can trap teams in rigidity.

But across domains, the data is precise. In a crisis, even experts regress to primitive heuristics. Stress degrades untrained intuition faster than it degrades protocol.

Cognitive armor does not eliminate instinct. It preserves the conditions in which instinct remains accurate.

As aviation psychologist Robert Helmreich put it: "Discipline makes intuition reliable."

The Power of the Pause

When the Apollo 13 service module exploded, mission control was awash in noise. Telemetry spiked. Screens flooded with contradictions. Panic was seconds away.

Flight Director Gene Kranz issued an order that baffled new controllers: "Let's take a quiet five."

· · ·

The room fell silent.

That five-minute pause preserved the mission. It restored orientation before the situation could worsen. Calm became the catalyst for survival.

Emergency management agencies are now institutionalizing similar behaviors—micro-breathing cycles, scripted communication, time-bounded resets. Calm is not a personality trait. It is an operational discipline.

Engineering Composure

Cognitive armor channels emotion rather than suppressing it. Fear, urgency, and uncertainty are forms of energy. Left unmanaged, they fragment thought. Harnessed, they sharpen perception.

Organizations can engineer composure through three design choices:

1. Treat stress as a signal, not a failure.

Rising tension becomes a cue to apply protocol, not a source of shame.

2. Build deliberate friction.

Decision windows. Verification steps. Structured pauses. These slow thoughts are just enough to restore clarity.

3. Train the pause.

Repetition converts reflection into reflex.

Briefly slowing down allows teams to go faster safely.

Cognitive Armor in the Age of Automation

As automation absorbs execution, human cognition becomes the sense-making layer. Yet machine speed amplifies stress. In security operations centers, analysts may review hundreds of AI-generated alerts per hour. False positives create exhaustion. Ambiguity creates tunnel vision.

Structured triage, micro-breaks, rotation protocols, and mental resets become critical. Automation accelerates complexity. Cognitive armor keeps humans from drowning in it.

The next frontier of competence will not be who knows the most, but who thinks clearly under distortion.

The Enduring Discipline

Cognitive armor is not resilience in the inspirational sense. It is the infrastructure of thinking under pressure. It turns chaos into choreography. It transforms fear into focus.

· · ·

Surgeons, pilots, traders, firefighters, investigators, security analysts—all rely on it whether they acknowledge it or not.

Composure is not innate. It is constructed. It must be rehearsed, reinforced, and renewed.

Like any armor, it must be inspected after every impact, polished through repetition, and strengthened by memory— especially the memory of what happens when it fails.

Chapter 15
The Architecture
of Attention

Every organization already has an architecture of attention. It is rarely intentional. It forms slowly, through dashboards, emails, metrics, and alerts, each built for function, not cognition. Together, they decide what people see, what they ignore, and eventually what they believe.

Attention is not infinite. It is rationed, budgeted, and biased. What gets visualized gets valued. What disappears from the screen often disappears from thought.

Design is not cosmetic. It shapes what people come to know. It shapes how reality is constructed inside an organization.

In one global company, the security operations center relied on AI tools to flag anomalies across a vast network. The system ranked alerts by threat likelihood and displayed thousands of events a day. Analysts, under pressure to keep up, learned to trust the ranking. Their eyes drifted to the "high probability" column because that was where attention felt rewarded.

Attackers watched the same pattern.

A sophisticated intrusion unfolded over several weeks in low-frequency, low-confidence events that sat below the threshold of urgency. The threat actor probed slowly, testing response boundaries. The signals were faint, but consistent.

No one saw them.

The design rewarded attention to the probable rather than the possible.

After the breach, the company did not simply add more alerts. It redesigned the entire view. Instead of static probabilities, the dashboard displayed rates of change and temporal relationships among events. Analysts were retrained to notice acceleration, not just height, and to ask a different question.

Not only "What is ranked highest?" but also, "What is evolving?"

. . .

Within months, true-positive detections improved, and false positives decreased. The interface had not only optimized the workflow. It had recalibrated perception.

That is the power of design. It decides what becomes visible, and in complex systems, what is visible becomes real.

Design as a Cognitive Lens

The same pattern appears far beyond cybersecurity.

In hospitals, electronic health records once showed vital signs as isolated numbers—blood pressure here, heart rate there, oxygen saturation in another box. The physician had to compute the trend while mentally juggling dozens of other tasks.

When one hospital system redesigned its interface to display vital signs as continuous trajectories, diagnostic accuracy improved. Doctors no longer saw a number. They saw a story.

In corporate risk dashboards, static red–yellow–green tiles encourage binary thinking. Safe or unsafe. On track or off. When those same dashboards were rebuilt to show trend lines and velocity, the conversation changed. Leaders began asking why risk was accelerating, not simply what color it was.

. . .

That shift looks subtle. It is not. It moves cognition from labeling to learning, from judgment to inquiry.

Design is a cognitive lens. It guides people in moving from data to meaning.

The Invisible Hand of Interface

Good design channels attention. Bad design hijacks it.

Every alert, chart, or metric competes for cognitive bandwidth. Too many, and everything becomes noise. Too few, and blind spots grow.

Humans are serial processors pretending to be parallel ones. A 2023 Stanford study found that workers exposed to more than 120 notifications per hour suffered a 30% drop in analytical accuracy. That is the rough equivalent of mild intoxication. Yet most digital workplaces still behave like slot machines for attention, not instruments of focus.

You cannot increase human attention. You can only direct it better.

Designing for discernment begins with accepting that biological limit.

. . .

Aviation: When Design Learned Humility

Aviation learned this before most fields.

In the 1950s and 1960s, investigators noticed many "pilot errors" looked strangely similar. The wrong switch. The wrong lever. The same misread gauge. Identical aircraft. Identical mistakes.

The conclusion was uncomfortable. The pilots were not failing the cockpit. The cockpit was failing the pilots.

Warning lights looked alike. Controls clustered without logic. Gauges were placed for wiring convenience, not for clarity under stress. In other words, the system engineered the error.

Human factors engineering was the response. Controls were grouped by function. Colors were standardized for urgency. Auditory alarms were reworked to be distinguishable even in chaos.

The cockpit ceased to be merely a control center. It became a cognitive habitat. Every dial and alarm became an ethical decision about how to make the right action easier to see.

. . .

What leaders once called "human error" was often a design failure in disguise.

Cockpits, Hospitals, and Command Rooms

The cockpit lessons were applied to medicine, nuclear energy, and emergency management.

In hospitals, redesigned infusion pumps now show clear confirmation prompts before drug delivery. That simple step has prevented overdoses that once traced back to ambiguous screens.

In nuclear plants, control panels were reorganized to group alarms by system, reducing cognitive overload when multiple failures cascade simultaneously.

In emergency command centers, map-based visualizations replaced long text lists of incidents. Responders can now see geographic clustering, anticipate spread, and stage resources before data catches up.

In every case, discernment was not only taught. It was engineered.

The Andon Cord for Knowledge Work

On factory floors, the andon cord gives any worker the right to stop the line when they see a defect. It is a visible, mechanical expression of psychological safety and quality control.

Knowledge work has its own version.

A software company was preparing a major feature launch for enterprise customers. Marketing had the press queued. Sales had reference calls booked. The go-live was set for Tuesday at 10:00 a.m.

On Monday night, a data scientist named Priya noticed something odd. One experimental cohort showed an uplift far beyond modeled expectations. It was not good news. It was implausible.

Two stories were competing for attention. The optimistic one said the feature was a breakthrough. The skeptic said the telemetry was incorrect.

At 6:00 a.m. Tuesday, the executive chat channel was already active. The CEO had posted a draft note celebrating the launch. Sales leaders were asking for green lights.

. . .

Priya opened the instrumentation repository. Two weeks earlier, a pull request had changed event names in the client library for a subset of platforms. During a rollback from a beta branch, some conversions were counted twice. The tests had passed. The change looked clean. The bug was subtle.

The company had an explicit andon policy for launches. Anyone could call a stop-ship if specific triggers were met. One trigger was an unexplained divergence between experimental cohorts exceeding a defined threshold.

Priya pulled the cord.

The default stop window was twenty-four hours. Leaders could override only with a written rationale in the same channel.

The mood tightened. The product manager asked for ten minutes. Engineering formed a small strike team with Priya and the telemetry owner. Within two hours, they reproduced the double count on a test device that moved between beta and stable branches. About twelve percent of the target population was affected.

Marketing paused press. Sales reframed reference calls around the product vision rather than immediate availability.

. . .

By noon, the team had a patch. Counts were confirmed clean. The launch moved to the next day. Beta customers received a short, honest note.

The AAR took fifteen minutes. The andon policy had done precisely what it was meant to do: protect the company from shipping under a false signal.

Leadership reinforced the behavior publicly. They praised the stop, documented the decision trail, and added a new control. Any telemetry schema change would automatically generate a diff for data science to review, along with a checklist of failure modes, before launch week.

The lesson is durable. In knowledge work, a visible right to stop the line, backed by clear triggers and a short clock, protects reputation and trust.

Put It To Work:

- Define objective andon triggers such as unexplained cohort divergence or abnormal metric spikes.
- Make the stop visible in the main launch channel with a timer and a named owner.
- Require written override rationales from leaders in that same channel.
- Auto-generate telemetry schema diffs and route them for checklist review.

- Run a fifteen-minute hot wash and lock in one permanent change.

Information as Terrain

In corporate life, the architecture of attention is often quiet but powerful.

A CFO's dashboard may highlight cost per unit and revenue per headcount. That design naturally nudges behavior toward cuts rather than experimentation. A sustainability team's board may focus on per-product emissions, which steers behavior toward compliance rather than innovation.

This is "metric myopia." Measurement replaces meaning. The design of the metric becomes the design of the moral story.

One multinational retailer discovered that its "on-time delivery" metric was creating perverse incentives. Suppliers shipped incomplete orders early to preserve their scores. Customers received partial shipments faster, but overall service was worse.

When the metric was redesigned to weight customer satisfaction alongside timeliness, behavior shifted within weeks. The architecture of attention had been corrected, and so had the ethics embedded in it.

· · ·

Design does not just control performance. It shapes conscience.

Scripts for Clarity Under Uncertainty

In turbulent moments, leaders are part of the interface as well. Their words become design elements in the architecture of attention.

Two simple scripts help keep attention aligned when information is incomplete.

Script 1: During ambiguity

"Here is what we know. Here is what we do not know. Here is what we are doing next. Your next update will be at [time]. If you need to decide before then, I will give you the tradeoffs in one paragraph."

Script 2: When extending the window

"We are extending the window by [time] to complete [two steps]. Risk to people is [low, medium, high]. Operations impact is [low, medium, high]. I own the next update at [time]."

These scripts do more than calm people. They format attention. They show where to look and when clarity will be refreshed.

Counterarguments: The Limits of Design

Skeptics are right about one thing. No interface can guarantee discernment.

Humans can still misread a clean display, ignore a critical chart, or cling to a narrative that contradicts the data. A well-structured dashboard will not fix a failing culture on its own.

But design determines where cognition begins. It is the architecture of first thought.

Bad design multiplies bias. A cluttered SOC screen that sprays false positives teaches analysts to ignore alerts. A medical alarm that fires constantly trains nurses to silence it reflexively. The medium trains the mind.

Good design does not remove bias. It gives people a better starting point for thinking.

Discernment is a human skill. It either grows or withers inside the environment that design creates.

Designing for Human Limits

Designing for discernment means building systems that

amplify human strengths and buffer human weaknesses. It is not about more data. It is about a better context.

Three principles stand out.

Limit signals to what matters. Every extra metric dilutes attention. Minimalism is not aesthetic. It is moral.

Emphasize narrative over volume. Show how events connect, not just how many exist. Humans understand stories more easily than scatterplots.

Surface uncertainty explicitly. A ninety percent confidence score should still look like a range, not a verdict. False precision is a design failure.

These ideas mirror what behavioral economists call choice architecture. Environments can guide decisions without coercing them. In a crisis, the same rule applies. Design often decides whether a human acts with clarity or confusion.

When Design Becomes Ethics

Every interface carries ethical weight.

In a hospital, one misplaced checkbox can cost a life. In an operations center, an overloaded screen can blind a team to a breach. On social platforms, engagement-maximizing feeds have amplified division and misinformation worldwide.

. . .

Designers, engineers, and leaders now share a new kind of responsibility, not for what people choose, but for what they can notice.

A well-designed system does more than display information. It cultivates discernment. It teaches its users how to see.

The same applies to every digital workspace. Every collapsed menu, default filter, and absent context window is a decision about visibility. Visibility is the currency of truth in complex organizations.

Discernment begins not with what we know, but with what we can see.

The next generation of intelligence systems will not be judged by how much data they hold. They will be judged by how clearly they help humans perceive.

The first step toward wisdom in any system, technical, social, or corporate, is demanding and straightforward.

You have to decide to design for it.

Chapter 16

Composure as a System

Composure is the quiet infrastructure of leadership. It is not serenity for its own sake. It is a method for maintaining clarity as the world accelerates. Where most people treat calm as a personality trait, resilient organizations treat it as a system.

They do not hope someone "stays cool under pressure." They build conditions where clarity is the default. In a real crisis, calm is not optional. It is oxygen.

The Discipline of Stillness

During the 2011 Fukushima Daiichi disaster, engineers faced a stack of failures that bordered on the unthinkable. Earthquake. Tsunami. Power loss. Meltdowns. Sensors failed, readings contradicted each other, and radiation climbed.

．．．

The teams that performed best were not simply the most technical. They were the ones who had trained for composure.

They used silent checkbacks, a structured pattern where every instruction is repeated, confirmed, and acknowledged before anyone moves. You see the same behavior in high-end kitchens and high-reliability operating rooms. It is a ritual that slows tempo on purpose. Command. Repeat. Confirm.

Those steps act as a cognitive brake. They pull attention away from fear and back to precision. In an environment where seconds felt critical, these teams chose to sacrifice raw speed for certainty.

The lesson is not that haste always kills. It is that haste blinds. Composure pulls perception back out of panic.

The Modern Crisis Reflex

Corporate crises play out differently, but the psychology is the same.

A data breach, a scandal, a cloud outage. Slack channels light up—email chains fork. Executives rush to issue statements. Everyone tries to fill the silence with activity.

· · ·

Motion is not coordination. Information without alignment produces confusion, not confidence. Each uncoordinated update that contradicts the previous one erodes credibility. Panic spreads faster than clarity.

The organizations that ride out these events best have learned to slow the clock. They named a crisis lead. They set explicit decision intervals. They centralize communication rather than letting each function create its own version of reality.

One technology firm learned this the hard way during a ransomware incident. Within hours, different groups had pursued conflicting paths. One team began payment negotiations. Another pushed toward a complete rebuild. Legal and communications sent inconsistent messages to regulators, customers, and staff.

The real damage came from that internal fragmentation.

Only when leadership consolidated command did the signal return; they imposed three-hour decision windows, required dual sign-off for anything external, and synchronized core teams with short, timed syncs. The attack did not become easier. The tempo did.

The crisis no longer dictated the rhythm. The team did.

Command of Tempo

Composure is not passivity. It is tempo control.

The first task of leadership in a crisis is to slow the loop just enough for sense to catch up with action. High-reliability fields do this on purpose.

In trauma care, teams call "stop moments" when information conflicts.

In military command, battle rhythm governs decision cycles, so units do not respond to every stimulus.

In aviation, sterile cockpit rules below 10,000 feet reduce distractions to allow attention to be focused.

These protocols institutionalize composure. They turn stillness into a procedure, not a mood.

The Science of Slow Thinking

Daniel Kahneman described two modes of cognition. System 1 is fast, intuitive, and emotional. System 2 is slow, deliberate, and analytical. Under stress, the brain defaults to System 1. That is how humans survived predators. It is also how organizations make confident, rapid mistakes.

. . .

Composure protocols are practical methods for returning to System 2 when it matters most.

Neuroscience backs this up. Teams trained in deliberate breathing and structured decision routines show reduced activity in the brain's alarm centers and increased activity in regions tied to planning and control. Calm is not mystical. It is mechanical.

You can train it as you would any other reflex. Repetition under pressure until the behavior appears automatically.

Structured Calm

Evidence from emergency medicine and command research points to the same conclusion. Structured calm improves performance.

A 10-second "micro-huddle" before a significant intervention in trauma rooms increased procedural accuracy. NATO studies of command teams showed that enforced decision intervals improved mission clarity without slowing operational response.

The principle is simple. Slow is smooth. Smooth is fast.

. . .

A composed organization does not move more slowly. It moves with less rework and fewer self-inflicted wounds. It can separate urgency from velocity. It can act with intent instead of impulse.

Thrust without a vector never ends well.

Building a Composure Culture

A culture of composure begins with design, not personality.

Four elements matter.

1. Decision windows

Instead of a permanent react mode, teams operate in defined cycles. The window might be three hours in a ransomware incident or fifteen minutes in a safety event. The point is rhythm. The window creates a shared expectation about when decisions will be made and when information will be refreshed.

That cadence prevents decisions from being made in isolation at the loudest moment in the room.

2. Challenge to first impressions

Under stress, the first explanation that feels coherent often becomes "the story," whether or not it is true.

To counter this, mature teams embed a second-question rule. Every initial assessment must be tested before it drives significant action. NASA does this through designated contrarians in flight reviews. Their job is not to be difficult. It is to make sure confidence has earned its footing.

This is structured skepticism, not cynicism.

3. Precision in language

Clarity under pressure depends on shared vocabulary.

Teams that use defined status codes, escalation tiers, and handoff formats can move quickly without talking past one another. This is why militaries use brevity codes and why emergency departments rely on structured handoff scripts.

Communication becomes part of the safety system, not an improvisation.

4. Leadership stillness

People mirror the emotional tone of the most senior person in the room. That is true in cockpits, operation centers, and boardrooms.

A visibly agitated leader amplifies anxiety. A composed leader sets an anchor. Their voice, body language, and timing become a reference point for everyone else's nervous system.

When stillness is modeled consistently, it becomes cultural gravity. Calm spreads as quickly as panic.

When Calm Is Misread

There is a predictable objection. Too much structure, critics say, creates bureaucracy. In fast markets or breaking crises, hesitation costs opportunity.

The key distinction is simple. Composure is not hesitation. It is disciplined pacing.

Look at flight operations on a carrier deck. Jets launch and recover on a moving runway surrounded by hazards: noise, speed, fuel, weapons. Yet every action is governed by slow, deliberate gestures and repeated confirmations.

The ritual of calm enables that level of speed without collision.

In business, the calmest teams often move the fastest over the whole arc of an event, because they are not burning time cleaning up preventable mistakes or revising contradictory messages.

Calm as a Cultural Asset

Composure is both method and message.

.   .   .

In every crisis, people watch leadership more than they read memos. They are measuring one question: "Are the people in charge still thinking?"

If the answer is yes, trust remains even when the news is bad. If the answer feels like no, the best data in the world will not restore confidence.

This is why calm cannot depend on one charismatic operator. When composure is personal, it leaves with the person. When it is built into routines, training, and shared language, it survives turnover and succession.

Calm becomes a reusable asset, not a lucky coincidence.

The Final Discipline

At its core, a composure protocol is straightforward.

- Slow down briefly before major decisions.
- Challenge the first story that feels right.
- Verify across roles and perspectives.
- Communicate with precise, repeatable language.
- Protect clarity as fiercely as you protect time.
- These are not soft skills. They are survival systems.

Composure is not the absence of fear. It is a command of tempo. Leaders who master it do not remove uncertainty. They make it navigable. Their calm is not a personality trait. It is an operational framework, an invisible structure that keeps the organization steady when everything around it starts to move.

Chapter 17
Cognitive Sovereignty

The defining battle of the information age is not for territory, commodities, or ideology. It is for attention. Every platform and algorithm competes to occupy mental space and secure a slice of the cognitive terrain. Engagement becomes influence. Retention becomes revenue. The modern economy is, at its core, a contest for the mind.

In that contest, autonomy is under pressure. The same tools that expand awareness can also erode the capacity to navigate it freely. Cognitive sovereignty—the ability to think independently under informational assault—is no longer a philosophical preference. It is the functional requirement for clarity in a world built to capture attention rather than respect it.

The Convenience Trap

Erosion rarely arrives through coercion. It comes through convenience.

Personalized feeds, predictive search results, and curated dashboards—each promises simplicity. But simplification has a cost. What appears as a service often becomes structured. Each suggestion narrows the horizon. Each recommendation shapes what you never see.

A 2023 MIT Media Lab study followed news consumers over six months. Participants primarily exposed to algorithmically curated content showed a measurable decline in willingness to consider opposing views. fMRI scans revealed a shift in neural reward patterns. Comfort came from confirmation; contradiction triggered stress. The effect was not ideological entrenchment. It was attentional conditioning. The brain began to crave validation and recoil from complexity.

The architecture of convenience had quietly rewritten cognition.

The New Front Lines

The consequences extend far beyond individual behavior. Modern conflict now unfolds inside cognition itself.

. . .

During Russia's 2022 invasion of Ukraine, the information battlespace became as dynamic as the physical one. Deepfakes of Ukrainian leaders, coordinated bot networks, and fabricated news outlets flooded channels with synthetic certainty. The goal was not persuasion. It was paralysis. When all sources feel compromised, the mind cannot anchor truth.

The countermeasure was not purely technical. They were distributed discernment.

Open-source intelligence groups—Bellingcat, GeoConfirmed, and OSINT Curious—assembled real-time verification networks across continents. Photographs were geolocated, videos timestamped, and claims tested by thousands of volunteers. Cognitive sovereignty became a collective act.

The lesson is blunt: national security now depends as much on media literacy as on military capability. The front line has moved from borders to bandwidth.

Corporate Cognitive Capture

The same forces shape corporate life in quieter ways.

Recommendation engines, decision dashboards, and automated workflows promise insight but often create dependence. When an algorithm mediates every judgment, the human sense-making loop dulls.

. . .

A 2022 Harvard Business Review study found that when senior managers relied heavily on AI-assisted systems, independent problem-solving dropped by nearly 30 percent. Many deferred to machine output even when it contradicted their professional judgment. Researchers called this *delegated cognition*—the outsourcing of thought to systems whose reasoning remains opaque.

Dashboards with color-coded confidence scores offer the illusion of clarity without revealing what the interface has filtered out. When visibility collapses, control collapses with it.

One global logistics firm confronted this after a series of dire forecasts driven by overly trusted analytics. Their fix was architectural: every dashboard now includes a "transparency panel" that shows model assumptions, data freshness, and blind spots. Decision quality improved, but the more profound shift was cultural: skepticism became normalized. Awareness replaced passive trust.

The Neuroeconomics of Capture

Neuroscience now frames attention as a finite biological resource. Every ping, every scroll, and every push notification taxes the same systems that regulate working memory and long-range planning. Digital environments optimized for maximum engagement do not preserve attention; they fragment it.

. . .

This fragmentation scales from personal fatigue to societal drift. A distracted public is easier to polarize and harder to mobilize. The line between economic exploitation and political manipulation thins.

Sovereignty is no longer just a civic concept. It is a cognitive one.

Resistance as Discipline

Cognitive sovereignty begins with boundaries—simple acts that reclaim mental bandwidth. Disabling notifications, scheduling offline hours, vetting sources, and curating information diets are not lifestyle choices. They are defensive tactics.

Two commitments define this discipline:

1. Awareness of architecture. Every interface is an argument about what matters. Every default sorts your attention before you can.

2. Intentional friction. A mind without friction is a mind shaped by whatever arrives first.

Leaders who schedule uninterrupted "quiet blocks" are not indulging themselves. They are protecting the last domain where strategic thinking still lives. Teams that disable alerts during crisis response are not ignoring the data. They are defending coherence.

Attention is a finite national resource. Its depletion is not just a productivity threat—it is a geopolitical vulnerability.

The Counterargument: The Case for Personalization

Personalization can strengthen autonomy when designed well. It reduces noise, increases relevance, and helps individuals focus on what they value.

The issue is not personalization itself. It is secrecy.

When a system filters without revealing its logic, agency evaporates. When it explains its logic, agency returns.

Spotify learned this when it introduced "Why You're Hearing This Song," a minor feature that shows which listening patterns triggered a recommendation. Trust increased—even when the recommendations were imperfect. The same principle should govern news feeds, corporate analytics, and AI systems: show the logic and you restore the choice.

Transparency is sovereignty.

Sovereignty by Design

Cognitive sovereignty must be engineered, not assumed. Modern systems require architectural guardrails:

- AI outputs must include explainability hooks.
- Recommendation algorithms must expose inputs and weightings.
- Dashboards must reveal what is filtered out, not just what is shown.
- Teams must be trained to recognize when fatigue, bias, or automation is steering their decisions.
- Cognitive security must evolve toward zero-assumption thinking: verify the source, question the interface, and challenge inherited frames.
- Organizations that thrive in the next decade will not simply manage information. They will defend attention integrity.

The Final Frontier

Nations once defended borders. Now they must defend bandwidth. Corporations once protected assets. Now they must protect judgment. Individuals once guarded their privacy. Now they must guard cognition.

The mind is the first and final frontier of freedom.

Cognitive sovereignty is not the rejection of technology. It is mastery within it. It is the discipline to question the frame, pause amid noise, and think through distortion.

In a world saturated with algorithmic noise, sovereignty becomes a form of courage—the ability to remain oriented when the informational weather turns against you.

That, more than any weapon or algorithm, is the new measure of strength.

Chapter 18

Navigating the Manipulated Environment

How Feeds, Search, and Politics Shape What We Think

Most people do not wake up inside a propaganda lab. They wake up inside a feed. The interfaces we rely on—news apps, social platforms, search engines—adapt to our habits and quietly reshape themselves around our preferences. For entertainment, that is mostly harmless. For news, judgment, and meaning, it becomes a subtle form of drift. The feed learns what holds attention and strengthens that pattern. Over weeks and months, the world begins to tilt.

This is not malice. It is mechanics. Algorithms are exquisitely effective at delivering more of whatever we already attend to. But cognition is porous. Repetition shapes familiarity, familiarity shapes comfort, and comfort shapes belief. A feed tuned for convenience becomes an architecture of influence.

The Mechanics of Drift

Personalized systems curate information the way a tailor shapes fabric—pulling in one direction, trimming in another. The result fits almost too well. It hides the seams.

Researchers who track online behavior consistently find the same pattern: people are gradually nudged toward material that mirrors their existing attitudes. Opposing views do not vanish; they reappear as provocations—weaponized snapshots stripped of nuance. Identity becomes the frame. Outrage becomes the fuel.

The danger is not that algorithms skew political opinion. It is that they have a narrow cognitive range. They condition people to expect simplicity and to distrust complexity.

Data Voids: The Ambush Zones

Search engines introduce another, more tactical vulnerability: the *data void*. A void forms when a topic spikes suddenly—an emerging slogan, a new conspiracy, a viral claim—and quality sources have not yet had time to publish. Manipulators anticipate these moments. They flood the gap with search-optimized posts, thin websites, and deceptive "explainers."

To a good-faith searcher, the results page looks plausible. But the landscape has been rigged. The first impressions—those that most strongly shape cognition—belong to whoever filled the void fastest.

A void is recognizable by its texture: unfamiliar sources on page one, thin citations, emotionally charged summaries, and claims without context. When that pattern appears, the correct move is procedural, not emotional. Widen the query. Add site limiters ("site:.gov", "site:.edu"). Search the organization behind the claim rather than the claim itself. Slow down. Step back. Let reputable sources catch up.

The first defense against manipulation is not knowledge. It is cadence.

The Firehose and the Familiar

Modern political messaging has adapted to this environment. Instead of constructing a single coherent argument, operators deploy many variations of the same message across as many channels as possible. RAND termed this the *firehose of falsehood*: speed over accuracy, volume over coherence, familiarity over proof.

The strategy exploits a basic human vulnerability: the illusory truth effect. A statement repeated often enough begins to feel true — not through evidence, but through erosion. Familiarity replaces verification.

· · ·

Awareness of this psychological quirk is a defensive weapon. When a claim sounds "true" simply because you have seen it everywhere, that is the moment to pause. Not because the claim is necessarily false, but because your recognition is no longer a measure of its accuracy. Familiarity is a signal to check, not a license to believe.

Cognitive Safety in Practice

Cognitive security does not require a Ph.D. It requires disciplined habits—the intellectual equivalent of hand hygiene.

Two-source verification: Read two credible sources with different editorial leanings. If the core facts match, you are on firmer ground.

Reverse search the source: If a claim is unfamiliar, search for the organization or person behind it. Legitimacy leaves a trail.

If a story is only minutes old, the informational environment is unstable; early impressions are most vulnerable to manipulation.

Switch modalities: If the search begins to feel like a rabbit hole—new jargon, anonymous sites, sudden emotional spikes

—slow down. Change your query. Move to established outlets. Look for primary documents.

Treat urgency as a red flag: Most of the manipulative information demands speed. The safest response is to delay. Waiting is a decision.

These habits accumulate. They do not remove noise, but they prevent noise from colonizing belief.

The Strategic Goal

The goal is not to escape the feed. Total disconnection is neither possible nor desirable. The goal is to prevent the feed from quietly shaping the boundaries of thought—what you consider, what you discard, what you trust before checking.

Cognitive sovereignty is built not through isolation but through discipline. Every structured pause, every cross-check, every decision to broaden context rather than narrow it strengthens the mental perimeter. Autonomy is reclaimed not in grand gestures but in small refusals.

It is a quiet form of resistance: the determination to think on your own terms in an environment designed to feel for you.

Chapter 19
Building Cognitive Immunity

Simple Habits, Better Tools, and What Schools Are Teaching

We will not resolve the modern information crisis one correction at a time. The volume is too high, the systems are too fast, and the incentives are too misaligned. What works— what has always worked — in high-noise environments is layering small defenses that build resilience. Clarity, in this era, is an ecosystem. It is built from habits, tools, and infrastructure that operate before confusion arrives, not after.

The emerging field calls this cognitive immunity: light, repeatable practices and structural safeguards that make individuals, teams, and institutions harder to deceive.

Cognitive immunity is not censorship, nor is it skepticism in extremis. It is a discipline of verification—gentle friction that keeps perception aligned with reality.

Credentialed Reality: The New Provenance Layer

One of the most promising shifts in modern information integrity is the rise of Content Credentials. This tamper-evident provenance label travels with images, video, or audio from the moment they are created. Built on the C2PA standard, the credential records who created the media, the edits made, and how the file has changed over time.

Think of it as a supply chain for truth—a chain of custody for pixels.

Credentials do not guarantee accuracy. They do something quieter and more crucial: they make manipulation visible.

Visibility forces honesty. It builds trust in authentic materials and creates friction with synthetic materials.

As more newsrooms, platforms, and camera manufacturers adopt C2PA, provenance becomes a first line of cognitive defense. If you publish media at work—internal memos, investigation footage, training videos—turn Content Credentials on by default. You are not just protecting your product; you are protecting your reputation's future.

· · ·

Provenance is clarity encoded at the point of creation.

Prebunking: Training the Mind Before the Attack

Correction is slow. Manipulation is fast. The most effective countermeasure is not fact-checking or policing content, but pre-inoculation—prebunking.

Researchers at Cambridge and Google's Jigsaw found that simple 90-second explainers teaching common manipulation tactics—scapegoating, emotional hijacking, false dichotomies —improved detection accuracy across large populations. Much like vaccines, prebunks prepare the cognitive immune system. People recognize the tactic when it appears, even if they've never seen the specific claim.

Accuracy prompts work the same way. In studies by Gordon Pennycook and colleagues, asking a single question—*"How accuracy do you think this is?"*—significantly reduced sharing of false headlines. Not because the prompt supplied new information, but because it restored attention to truthfulness, shifting cognition from reflex to reflection.

Small friction scales. So do small reminders:

- A one-line banner in a group chat: *"Verify high-stakes claims before forwarding."*
- A standing meeting opener: *"What would falsify this?"*

- A Slack bot that asks *"Are you sure?"* on posts with flagged keywords.

These micro-checks accumulate into organizational stability.

Deepfakes: Triage Before Trust

Deepfakes deserve their own doctrine. Detection systems are improving—thanks in part to pioneers like Hany Farid—but no filter will ever be perfect. The adversary continually adapts. Reliance on detectors alone creates a new fragility: false confidence.

The rule is simple: Treat detectors as triage, not verdict.

When a shocking clip appears:

- **Check provenance**—look for Content Credentials or source identifiers.
- **Pull a still frame** and run a reverse-image search.
- **Search for authoritative confirmation** from reputable outlets.
- **Apply temporal skepticism**—new claims are the least stable.

If you cannot authenticate, delay amplification. Waiting is not indecision; it is responsible command of tempo. In high-noise environments, a pause is a tactical choice.

• • •

Organizations should set explicit rules for when to remove, flag, or escalate suspicious media content. Clarity must be procedural, not improvisational.

The Long Game: Education as Infrastructure

Cognitive immunity cannot be built on tools alone. It must be taught—early, systematically, and without ideology. Some nations already understand this.

Finland integrates media-checking habits into everyday coursework and offers a public "Elements of AI" course accessible to non-engineers.

Norway and the Nordic systems embed digital citizenship into routine schoolwork.

Singapore teaches "Cyber Wellness" and runs a national "Digital Skills for Life" program.

Japan provides practical guidelines for using generative AI in K–12 classrooms.

Taiwan has built a whole-of-society model: teacher training, media-literacy events, and public fact-checking drills.

Their collective lesson is simple: Habits taught early become reflexes when it matters most.

Critical evaluation cannot be an elective. It must be cultural muscle memory.

A Family Protocol: The One-Minute Double Check

If you want a civilian version of crisis tradecraft—a daily routine that families or teams can adopt—start with this:

When a claim could affect someone's safety, reputation, or rights, pause for one minute. In that minute, find two independent sources that agree on the core facts. If you cannot see them, do not amplify.

This single protocol neutralizes most viral manipulation. Add one more habit: correct your own mistakes publicly and promptly. Credibility is not earned by perfection but by visible accountability.

Keep a folder of trusted reference sources—a small, curated cognitive perimeter. The tighter the perimeter, the clearer the mind.

The Quiet Strength of Small Habits

Cognitive immunity is not a grand innovation. It is a layering of simple habits: provenance, prebunking, verification, intentional friction, and early education. None is dramatic. All are powerful.

Because the fight for clarity in the modern era does not hinge on any one tactic—it hinges on accumulation. Small disciplines practiced by many people become an immune system. Small frictions prevent large cascades. Small pauses restore sovereign thought.

The solution is not heroic skepticism. It is the quiet, repeatable practice of seeing truth on purpose.

Chapter 20
Debate as Operational Tradecraft

A Plain Plan That Works for Any Team

In a recent ASIS *Security Management* article, I made a simple claim: **debate is not a performance—it is a working method built for the AI age.** In an environment where models, dashboards, and feeds generate fluent answers on command, teams need a discipline that restores clarity, structure, and accountability. Debate provides precisely that.

At its core, debate forces a team to do four things modern systems often obscure: make one claim, show the evidence, name the assumptions, and face the counter-case before acting. Those habits sound basic. Under pressure, they become rare. They anchor cognition when information moves faster than instinct, and they give leaders an audit trail for why a decision was made.

. . .

This chapter turns that article into a room-temperature operating model—a format any team can run with a whiteboard, a timer, and fifteen minutes of focus. The goal is not point-scoring. It is cognitive alignment under conditions that commonly distort perception.

Debate, reframed this way, becomes muscle memory for clarity.

The Debate Protocol: A Practical Clarity Engine

Start with a scenario that feels real:

"A video of our executive asking to change payroll accounts is spreading. Is it real?"

"A hashtag tied to our company is trending after a protest. Do we respond now, later, or not at all?"

"The model output says risk is elevated in Region X. What action do we take?"

Give the team five minutes to gather sources that *anyone else in the room can check later.* Screenshots and URLs get saved. Emotion does not count as evidence. This teaches source hygiene, not speed.

Then run the three roles:

1. The Advocate

States the recommended action in one sentence. Presents the evidence in order. Declares a confidence level in plain language—*possible, likely, very likely.* This forces precision, not performance.

2. The Challenger

Probes the weak points without theatrics. Challenge assumptions, missing context, and alternative interpretations. Their job is not to win. It is to strengthen the argument by testing it.

3. The Mapper

Charts what both sides agree on, what remains unknown, and what the next sensible step is *right now.* This role turns adversarial tension into alignment.

You close the drill with a short after-action card from each participant:

- The strongest argument they still disagree with and why,
- One bias they noticed in their own thinking,
- What new information would change their mind?

This last step builds a culture where updating your view is a strength, not a weakness. It normalizes epistemic humility—an essential trait when models, feeds, and interfaces can generate confidence far faster than truth.

· · ·

Run this weekly, and something predictable happens: Teams stop speaking in vibes and start talking in reason. Leaders hear uncertainty earlier instead of when it's too late. Bias gets surfaced, not buried. Action becomes proportional, not reactive.

This is operational skepticism without paralysis.

Onboarding the Non-Technical: Portable Habits

Not everyone comes with technical fluency, and that is fine. Cognitive security does not require engineers—it requires *habits*.

Give non-technical teammates a simple four-step check before they forward or respond to anything high-stakes:

1. **Slow down.**
2. **Look up who made the content.**
3. **Check the claim across two independent outlets.**
4. **Share only if it helps someone, not just because it is interesting.**
5. Add two bookmarks to their browser:
6. One for image and video checks (reverse-image, Content Credentials viewers) and one for a trusted fact library.

Then pair a tech-comfortable analyst with an experienced operator. The analyst brings tools; the operator brings context. The result is a judgment that neither could produce alone.

Start with low-stakes examples so the skill becomes a reflex rather than a source of performance anxiety.

Synthetic Pressure Drills: Preparing for Deepfakes and Rapid Claims

Deepfakes are now moving into live calls, quick-hit scams, and in-the-moment manipulation. Teams need to practice *inside* the environment where deception will occur.

Run a "CEO fraud" scenario using a polished fake video or audio clip. Train the team to:

- reverse-search frames,
- Check for Content Credentials,
- Verify through known channels instead of the number presented,
- Identify urgency as a red flag, not a command.
- Discuss the limitations of detectors openly. Reinforce that the right stance is trust but verify—always.

Watching teams deploy debate skills—clean claims, explicit assumptions, checkable evidence—under the psychological pressure of a convincing fake builds confidence for real-world velocity.

Why This Works: The Cognitive Mechanics

Teams that practice this see a shift within weeks:

- Analysts speak more clearly about what they know and what they do not.
- Operators find that the process accelerates real work instead of slowing it.
- Leaders get early warning of uncertainty rather than late surprises.

You end up with decision environments where:

- Confidence is earned,
- Evidence is organized,
- Assumptions are visible,
- Disagreement is structured,
- Updates are normal,
- And action remains proportional to truth, not tempo.

In the age of synthetic answers, cognitive pressure, and identity-driven feeds, this is an edge.

Debate—properly run—is collaborative skepticism in service of action. It does not make teams slower. It makes them clearer. It turns hesitation into structure and urgency into alignment.

. . .

When reality becomes fluid and signals fragment, the advantage belongs to teams that can test stories, hold doubt without freezing, and move with proportion.

This is tradecraft for the modern age. And you can teach it in any room.

Afterword

The Architecture of the Future

We are living inside a permanent storm. Signals arrive faster than judgment. Feeds multiply faster than meaning. The world's nervous system—sensors, models, dashboards, alerts —is expanding whether we shape it or not. And like any nervous system, it learns what we reward, forgets what we ignore, and amplifies what we build into its architecture.

The real design problem of our era is simple: what do we choose to amplify?

Information is infinite. Attention is not. Architecture decides which one wins.

Security in this new terrain is not a wall you build once. It is a rhythm you maintain. The work is calibration—finding the line between vigilance and overreach, between clarity and overload, between insight and intrusion. Every dashboard, every filter, every policy is a note in a larger composition that only becomes coherent when human judgment brings the system into tune.

We have learned the hard way that data does not guarantee wisdom.

Speed does not ensure safety.

Automation does not absolve responsibility—it magnifies it.

The more intelligent our systems become, the more deliberate we must be.

The First Principles That Still Matter

To navigate a world where information expands exponentially, we need disciplines that predate every algorithm on earth:

Curiosity—the willingness to ask better questions than the system can answer.

Humility—the instinct to seek disconfirming evidence when the result feels too clean.

Discernment—the discipline to separate loud signals from strong ones.

These are not sentimental virtues. They are operational necessities.

A team that can think critically under pressure will consistently outperform a team that merely reacts faster. A leader who can pause for clarity will outmaneuver one who races toward certainty.

Cognitive sovereignty—the ability to stay self-directed in an age of engineered attention—may become the most valuable security control of all.

The Symbiotic Turn

There is reason for optimism, not just caution.

The same technologies that flood us with noise also carry the potential to extend human insight.

Machine learning can identify subtle patterns that save lives.

Sensors can stabilize supply chains and ecosystems in real time.

Autonomous systems can scale decisions we once treated as impossible.

The challenge is not capability.

It is an intention.

The next evolution of intelligence will not be artificial; it will be symbiotic.

We are beginning to treat our systems not as replacements for human judgment, but as partners—interfaces that accelerate awareness while preserving agency.

The boundaries between sensing and interpretation, between prediction and understanding, are becoming porous. The organizations that thrive will be those that cultivate:

- Technical literacy to see how systems compute, and
- Moral literacy to decide how those systems ought to behave.

Symbiosis is not efficiency—it is alignment.

Choosing What We Become

Hope rests on the idea that we can build systems that reflect not only our logic but also our ethics. Tools that extend our perception without eroding our autonomy. Architectures that reward clarity over manipulation, curiosity over complacency.

This book has argued for a new posture—one where intelligence is adaptive, distributed, and principled. One where leaders become architects of attention, not merely executives of output. One where discernment is trained the same way we train technical skills. One where calm is designed, not discovered.

In that sense, every practitioner, analyst, and leader now holds the same quiet responsibility:

To decide what the system sees, how it learns, and what it amplifies.

Noise will continue to multiply.

Chaos will continue to accelerate.

Both are facts of the modern environment, not failures.

The challenge is to hear through the static—to find the patterns that matter and the signals that guide action. The advantage belongs to the teams that can build processes that adapt faster than the turbulence around them.

Security, as ever, is not a product.

It is a process.

The future will reward those who remember that—and build accordingly.

Epilogue

The Quiet Horizon

If there is one lesson from this age of acceleration, it is that clarity is earned and not handed down, not automated, not guaranteed. We have never been able to observe more of the world, and yet we have rarely felt less sure about what any of it means. The tools that illuminate also overwhelm. The systems that extend our reach can just as easily narrow our judgment. We can measure almost everything — but meaning remains a human responsibility.

The years ahead will bring more synthetic realities, more automated confidence, and more invitations to outsource our attention for convenience. Some will drift with the current. Others will adapt with intent. The divide will not be technological; it will be cognitive. The advantage will belong to the people and organizations that treat awareness not as a stream of inputs but as a discipline.

215

. . .

Those who cultivate calm amid confusion will define the next century. Those who can hold ambiguity without panic will make the right calls when the stakes peak. Those who understand that discernment — the ability to decide what *not* to attend to — is a form of power will remain self-directed in an era engineered for distraction.

Our task is not to predict the future but to prepare for its tempo. It will be fast. It will be saturated with competing truths. It will confront us with floods of information that appear authoritative, urgent, and coherent — even when they are none of those things. But it will remain human, because we are still the ones who assign meaning, steward trust, and determine consequence. We remain the center of the loop.

Somewhere ahead, between automation and autonomy, lies a new equilibrium — a partnership where systems amplify human judgment rather than replace it, where intelligence becomes symbiotic rather than adversarial, where speed serves clarity rather than erasing it.

That is the quiet horizon. Not a utopia. Not a collapse. Just the ongoing work of learning how to listen again, how to question our assumptions, how to design our own attention rather than surrender it.

. . .

If we can do that — if we can keep returning to curiosity, humility, and deliberate perception — then the signal will always find its way through the noise.

And so will we.

Author's Note

Author's Note

Signals in the Noise began as a field manual and became a meditation. What started as notes on threat detection, cognitive load, and situational awareness evolved into a broader inquiry into how humans think under pressure and what it means to remain human when the systems around us begin to think, too.

I wrote this book from the intersection of two worlds: one defined by physical security, risk, and operational discipline; the other by data, intelligence, and machine learning. In both cases, the same principle applies: clarity comes from the process. The most resilient organizations and the most grounded individuals treat awareness as a verb, not a state.

A simple conviction shaped the essays here: technology should enhance discernment, not replace it. In a century dominated by speed and simulation, our advantage will come not from having more information, but from understanding what it means.

Ice Station Zebra remains my workshop for that idea. What began as a design studio for coins and symbols has grown into a philosophy about attention, the artifacts we carry, the stories we tell, and the signals we leave behind. The Reed Group extends that same ethos to systems and strategy: how organizations build clarity amid complexity, translate principles into practice, and design for resilience rather than reaction.

Signals in the Noise is only the first volume in that larger pursuit. The subsequent explorations, Singularity, Private Intel, and Behavioral Barriers, will trace how intelligence evolves as it becomes more private, more synthetic, and more behavioral. Each will follow the same arc: from the tactical to the philosophical, from the immediate to the enduring.

I remain cautiously hopeful. The tools we are building today could easily amplify division or deepen understanding. The difference will depend on how consciously we create them and how seriously we take the responsibility of staying awake.

If this book leaves you with one idea, let it be this: the signal is still out there. It always was. Our job is to learn how to listen.

— Timothy E. Reed

www.ingramcontent.com/pod-product-compliance
Lightning Source LLC
Chambersburg PA
CBHW071556210326
41597CB00019B/3275